The Ciba Foundation is an international scientific and educational charity. It was established in 1947 by the Swiss chemical and pharmaceutical company of CIBA Limited—now CIBA-GEIGY Limited. The Foundation operates independently in London under English trust law.

The Ciba Foundation exists to promote international cooperation in biological, medical and chemical research. It organizes about eight international multidisciplinary symposia each year on topics that seem ready for discussion by a small group of research workers. The papers and discussions are published in the Ciba Foundation symposium series. The Foundation also holds many shorter meetings (not published), organized by the Foundation itself or by outside scientific organizations. The staff always welcome suggestions for future meetings.

The Foundation's house at 41 Portland Place, London W1N 4BN, provides facilities for meetings of all kinds. Its Media Resource Service supplies information to journalists on all scientific and technological topics. The library, open seven days a week to any graduate in science or medicine, also provides information on scientific meetings throughout the world and answers general enquiries on biomedical and chemical subjects. Scientists from any part of the world may stay in the house during working visits to London.

Ciba Foundation Symposium 146

PHOTOSENSITIZING COMPOUNDS: THEIR CHEMISTRY, BIOLOGY AND CLINICAL USE

A Wiley-Interscience Publication

1989

JOHN WILEY & SONS

Chichester · New York · Brisbane · Toronto · Singapore

©Ciba
Publish
Suggest
Ciba Fc
Ciba Fc
viii + 24
Library
Photose
use.
Inc
ISE
1. I
Cong
Photo
RM6
615.8
British
Photose
and c
1. Ch
I. Bo
founc
540
ISBN
Phototy
Printed

G. Jori De
Padua, It
H. Kato D
Shinjuku
A. H. Ka
Parkville
D. Kessel
Medicine
H. W. Lin
Medicine
A. J. MacF
4BS, UK
J. Moan
Norwegi
3, Norw
A. R. Mor
43606, U
A. Rook
Medicin
K. Schiew
Switzerl
T. G. Trus
Street, F
S. Ullrich
Andersc
USA
H. van de
Fédéral
J. E. van I
of Sher
B. C. Wil
711 Co

randomized comparative studies of the safety and efficacy of the drug. One must show that the drug is effective by comparison with the standard therapy. However, we are often working in a group of patients for whom there is no longer any standard therapy.

The first lung cancer trial compares PDT with Photofrin II and thermal ablation therapy using an Nd-YAG laser, the standard treatment for patients with obstructing or partially obstructing bronchial carcinoma. Patients with relatively advanced stages of lung cancer are being studied in this trial. The purpose of treatment is to provide palliation; there is no expectation that patients in whom disease is so far advanced are likely to be cured by this procedure. The endpoints are the effectiveness of the palliation and how long it lasts. A second trial is being done on patients in earlier stages of lung cancer who have not yet received radiation therapy. Patients in relatively early stages of disease may have extrabronchial tumours that are not affected by an endobronchial treatment by PDT or YAG laser. Therefore treatment is combined with X-irradiation, which is more extensive and reaches some of the nodes on the outside of the bronchi. This combined treatment is compared with the standard radiation therapy. Most of these patients would not be surgical candidates, for a variety of reasons.

In Phase III trials on cancer of the oesophagus patients are also being treated for palliative purposes, being in relatively late stages of disease. The standard thermal ablation therapy of an obstructing tumour in the oesophagus with a YAG laser is being compared with PDT in the first trial. The second oesophageal cancer trial is a single-arm study. The patients all have completely obstructing tumours, defined by inability to pass a guide wire through the oesophagus—a very stringent criterion. The FDA feel that the YAG laser is not safe in these patients. That allows us to do only a single-arm study, in the absence of an acceptable control.

The Phase III trial on bladder cancer is a randomized trial of prophylactic treatment of transitional cell carcinoma with single or multiple tumours in the bladder, compared with no treatment. The tumours are removed, usually by resection, and patients are treated with photodynamic therapy in an attempt to prevent recurrence of disease. There is not much precedent for this study. To my knowledge, PDT has never been shown to be a prophylactic treatment. So this trial has been controversial, because PDT has generally been expected to be most useful against carcinoma-in-situ, which is a difficult disease to treat in the urinary bladder, frequently requiring radical cystectomy. A protocol is now being formulated for treatment of the earlier stage carcinoma-in-situ.

That these Phase III trials are being carried out in the USA for regulatory approval doesn't mean that they are the best applications of PDT. They have been selected for their apparent efficacy, on the basis of earlier clinical information, and for what was thought to be ease of completion of the studies. Forty-five centres and 90 physicians are involved, with obvious attendant logistic problems. It is a slow process, but changes in regulatory procedures in the USA may mean that these trials will be shortened. At present the trials are expected to require 18 months for completion.

Regulatory approval in Canada and in Europe could be granted within the year, if the authorities there accept data from the Phase II trials that have already been completed, particularly those for lung cancer. Data from 200 lung cancer patients have been retrospectively collected and collated at three centres in the USA (Oscar Balchum at the University of Southern California, Jim McCaughan at Grant Hospital, Columbus, Ohio and Ann-Marie Regal at Roswell Park). We hope these Phase II trials will provide a good indication of safety and possibly of efficacy. Clinical trials have also been initiated recently in Japan. In China, many physicians are already using photodynamic therapy.

In this symposium we shall discuss basic science and medicine, but this is not very worthwhile unless PDT becomes available to clinicians. It has always disappointed me that, although we have been treating patients since 1976 at Roswell Park and nobody questions the effectiveness of the treatment, PDT is still not approved. Regulatory trials are only done if a company is convinced that a lot of money can be made from the therapy. The average cost of such trials in the USA is $100 million and the procedures take about ten years. An enormous amount of money, time and energy is spent on regulatory issues which are almost unrelated to medicine. That is how the system works, and in my view it is wrong. The FDA have made changes in the regulations to allow some drugs to be approved which may be useful for a small group of patients, but most drug companies don't touch those things.

These issues should be borne in mind as the background to our discussions in this symposium on the basic science and clinical use of photosensitizing compounds.

What are the ideal photoproperties for a sensitizer?

A. J. MacRobert, S. G. Bown* and D. Phillips

The Royal Institution of Great Britain, 21 Albemarle Street, London W1X 4BS, UK, and
**The National Medical Laser Centre, University College, London, UK*

Abstract. For effective photosensitization of malignant tumours, a sensitizer should exhibit appreciable absorption at red to near-infrared wavelengths and generate cytotoxic species via oxygen-dependent photochemical reactions. These photosensitization mechanisms rely on the excitation of the sensitizer from its electronic ground state to the fluorescent singlet state, which is in turn transformed into the longer-lived triplet state. Efficient formation of this metastable state is required because it is the interaction of the triplet state with tissue components that generates cytotoxic species such as singlet oxygen. Measurement of fluorescence provides a useful means of probing the sensitizer, particularly because detection sensitivities for both the triplet state and the cytotoxic singlet oxygen are much lower. With the development of chemically well-defined sensitizers, *in vivo* fluorescence detection has the potential to provide quantitative assessment of photoactive sensitizer distributions. The reactive properties of sensitizer triplet states and species such as singlet oxygen can result in significant sensitizer photodegradation, which may appear undesirable but can be exploited *in vivo* under certain conditions.

1989 Photosensitizing Compounds: their Chemistry, Biology and Clinical Use. Wiley, Chichester (Ciba Foundation Symposium 146) p 4–16

The photophysical and photochemical properties of a photosensitizer are collectively known as 'photoproperties', upon which the photosensitization efficiency or photocytotoxicity is dependent. The assessment of suitable photoproperties for a model sensitizer is not straightforward because the heterogeneous nature of biological systems can profoundly affect these properties. For example, the physical state of the sensitizer when taken up by cells is a major factor because generally there are large differences between the photoproperties of an aggregated species (e.g. dimer) and the monomer. In this paper we shall attempt to give some insight into the importance of these interactions in relation to photoproperties, but firstly we shall review the mechanisms of photosensitization.

On absorbing light of the appropriate wavelength, the sensitizer is converted from a stable electronic structure (S_0, the electronic ground state) to an excited state known as the singlet state (S_1*), which is short lived and may undergo conversion to a longer-lived excited state known as the triplet state (T_1*), which is the photoactive state responsible for the photochemical generation of cytotoxic species. A number of competing processes, such as fluorescence decay of the singlet state, may also occur (Equations 1–5).

$S_0 + h\nu \rightarrow S_1^*$	absorption	(1)
$S_1^* \rightarrow S_0 + h\nu$	fluorescence	(2)
$S_1^* \rightarrow S_0 + \text{heat}$	internal conversion	(3)
$S_1^* + M \rightarrow S_0 + M$	fluorescence quenching	(4)
$S_1^* \rightarrow T_1^*$	intersystem crossing	(5)

The lifetime of the singlet state is generally less than 1 μs and the main role of this state in the photosensitization mechanism is to act as a precursor of the metastable triplet state. However, its involvement cannot be overlooked because if the fluorescence is quenched through binding of the sensitizer to a particular substrate (Equation 4) the overall excitation efficiency from the ground to triplet state is correspondingly reduced. This excitation efficiency is defined as the triplet state quantum yield (Φ_T; probability of triplet state formation per photon absorbed), which for photosensitizers should ideally approach unity. Interaction of the metastable triplet state (which in de-aerated solutions has a lifetime extending to the millisecond range) with tissue components may proceed via either a type I or II mechanism (or a combination). A type I process can involve hydrogen abstraction from the sensitizer to produce free radicals or electron transfer resulting in ion formation. The type II mechanism, in contrast, exclusively involves interaction between molecular oxygen and the triplet state to form an electronically excited state of O_2 known as *singlet oxygen* ($^1O_2$*), which is highly reactive in biological systems. It is widely thought that the type II mechanism underlies the oxygen-dependent photocytotoxicity of sensitizers used for photodynamic therapy (PDT).

$$\text{Type II:} \quad T_1^* + O_2 \rightarrow S_0 + {}^1O_2^*$$
$$ {}^1O_2^* + R \rightarrow \text{photo-oxidation}$$

The near-resonant energy transfer from the triplet state to O_2 can be highly efficient: the singlet oxygen yield may approach the triplet state yield. An important feature of this mechanism which is sometimes overlooked is that when the sensitizer transfers electronic energy to O_2 it returns to its ground state. Thus the cytotoxic singlet oxygen species is generated without chemical transformation of the sensitizer, which may then absorb another photon and repeat the cycle. Effectively, a single photosensitizer molecule is capable of generating many times its own concentration of singlet oxygen, which is clearly a very efficient means of photosensitization provided the oxygen supply is adequate. Experimental evidence for the predominance of the type II mechanism is based on *in vitro* studies of cellular photosensitization; singlet oxygen scavengers are found to suppress the photocytotoxic effect (Valenzo 1987). In the following section experimental methods are briefly discussed and data for several promising photosensitizers are presented. In the discussion section aspects of a number of photoproperties (absorption, singlet oxygen generation, fluorescence and photostability) are specifically addressed.

Experimental methods and data

Spectroscopic techniques used to determine the photophysics and photochemistry of sensitizers are reviewed in Phillips (1989). Studies of the dynamics of the excited states rely on time-resolved measurements of fluorescence and absorption, which have been greatly facilitated by the development of a variety of pulsed laser techniques. For example, laser flash photolysis has been used to probe sensitizer triplet-state kinetics in model systems (e.g. protein-bound sensitizer) and in living cells cultured *in vitro* (Firey et al 1988, Truscott et al 1988). Time-resolved fluorescence detection is used to study the dependence of the properties of excited singlet states on environmental factors (e.g. different solvents) and the investigations are now being extended to *in vitro* systems. Detection of singlet oxygen is, however, much more challenging experimentally because this species may be rapidly scavenged *in vitro* through both physical quenching and chemical reactions which keep the $^1O_2{}^*$ concentrations below the detection levels of available techniques (Firey et al 1988). In solutions of sensitizers in deuterated solvents, singlet O_2 is readily detectable because of the reduced rate of quenching by these solvents (e.g. D_2O). Steady-state studies also generate useful data on fluorescence, triplet-state and singlet oxygen quantum yields using comparative methods. Table 1 lists some of the photoproperties of a number of promising photosensitizers together with data for the haematoporphyrin derivative (HpD)/dihaematoporphyrin ether (DHE) sensitizer which is already in clinical use. Peak extinction coefficients are given only for the red wavelength spectral region and, with the exception of DHE, the data refer only to the monomeric form of the sensitizer dissolved in a suitable solvent.

Discussion

Absorption

For PDT strong absorption is desirable in the red and near-infrared spectral region for which tissue is relatively transparent. Another advantage of red wavelength irradiation is that the potential mutagenic effects encountered with UV-excited sensitizers are avoided. Ideally, peak absorption of the sensitizer

TABLE 1 Sensitizer photoproperties: experimental data

Sensitizer[a]	*Extinction coefficient (M^{-1} cm^{-1})*	ϕ_F	ϕ_T	$\phi(^1O_2)$	*Reference*
DHE	5×10^3 (630 nm)	<0.1	—	0.25	Keir et al (1987)
m-THPP	5×10^3 (648 nm)	0.12	0.69	0.57	Bonnett et al (1988)
$AlPcS_4$	1.7×10^5 (675 nm)	0.52	0.42	0.3	Darwent et al (1982) Keir et al (1987)
ZnET2	3×10^4 (660 nm)	<0.1	0.84	0.6	Borland et al (1988)
Bchla	9×10^4 (780 nm)	—	0.3	0.3	Borland et al (1988)

[a]DHE, polyhaematoporphyrin ester/ether; *m*-THPP, tetra(*meta*-hydroxyphenyl)porphyrin; $AlPcS_4$, aluminium phthalocyanine tetrasulphonate; ZnET2, zinc aetio-purpurin; Bchla, bacterio-chlorophyll *a*.

should be near 1000 nm, on the grounds of optimum penetration depth in tissue. However, although there are some dyes which absorb from 800–1100 nm, the generation of singlet oxygen via the type II mechanism is energetically unfavourable because the triplet-state energies of these dyes are too low. It has been suggested that use of two-photon infrared excitation (equivalent to visible single photon excitation) would circumvent this problem, but, despite the availability of very reliable pulsed Nd-YAG lasers, the non-linear intensity dependence of multiphoton absorption, which is an inherently weak process, would only introduce further complications. To minimize skin photosensitization by solar radiation, the sensitizer's absorption spectrum should consist of a narrow red wavelength band, with little absorption at other wavelengths down to 400 nm, below which solar irradiation falls off steeply. Phthalocyanine dyes fulfil these criteria, unlike porphyrins (Tralau et al 1989).

From Table 1 it can be seen that, apart from *meta*-tetra(hydroxyphenyl)-porphyrin (*m*-THPP), the new photosensitizers all absorb more strongly at longer wavelengths than DHE, and *in vivo* experiments (studying sensitization with a mixture of sulphonated aluminium phthalocyanines (AlPcS) versus that with DHE, Bown et al 1986) have demonstrated that the photosensitization efficiency is significantly improved as a result of the higher penetration depth for 675 nm compared to 630 nm irradiation. Phthalocyanines are the strongest absorbers in this group; thus their use would be advantageous, provided the *in vivo* concentrations are below the level at which the strong photosensitizer absorption would limit the overall penetration depth of the light at the irradiation wavelength (Wilson et al 1986). A more subtle effect on photosensitization with a strongly absorbing dye is apparent when low repetition rate, high-pulse energy

($\geqslant$1 mJ per pulse) laser excitation is used. High pulse energies can induce saturation or transient bleaching of the sensitizer during the laser pulse and consequently much of the energy supplied is not efficiently absorbed by the sensitizer. It has been suggested that this effect accounts for the lack of photosensitized damage in normal rat colon sensitized with AlPcS and irradiated by a 5 Hz, 25 mJ-per-pulse flash lamp-pumped dye laser (Barr et al 1989b).

Singlet oxygen generation

The singlet oxygen yields of the new sensitizers shown in Table 1 are all similar and greater than for DHE, although the value for DHE is an estimate based on the assumption that a molecule of DHE is composed of approximately five haematoporphyrin rings; intracellular hydrolysis may reduce this number, therefore $^1O_2{}^*$ generation may be more efficient. Singlet oxygen yields for aggregates of the other sensitizers in Table 1 would be significantly lower than those given for the monomeric form, owing to the low triplet-state yields of aggregates, and, with the possible exception of *m*-THPP, the absorption spectra of the aggregates would be very different. In short, aggregation has a deleterious effect on the sensitizer photoproperties, although whether this general statement also applies to the biological properties remains unclear. Two of the sensitizers, aluminium phthalocyanine tetrasulphonate ($AlPcS_4$) and zinc acetio-purpurin (ZnET2), incorporate central metals chosen to increase the triplet-state and singlet oxygen yields without significantly shortening the triplet-state lifetimes, which should ideally be greater than 10 μs (in de-aerated solutions) to allow efficient singlet oxygen generation in tissue. Substitution with certain metals, however, (for example copper, which is paramagnetic) can markedly reduce the triplet-state lifetime, which explains why sulphonated copper phthalocyanine (CuPcS) is inactive *in vitro* (Chan et al 1987).

Fluorescence

Fluorescence is an inherent photoproperty of a sensitizer and is of practical interest for PDT because fluorescence detection may be used to probe sensitizers *in vivo* using fibre-optic techniques. For the successful exploitation of PDT, we would argue that the sensitizer should be a good fluorophore because fluorescence detection has sensitivity far superior to that of other techniques. However, some confusion regarding the interpretation of fluorescence has arisen from the use of HpD sensitizers as a diagnostic technique to identify malignant tissue. These sensitizers are mixtures of compounds with diverse photosensitization and fluorescence properties (aggregates and monomeric species; Dougherty 1987) and are therefore not well-suited to quantitative detection of fluorescence. Moreover, it has been suggested that weakly fluorescent HpD aggregates may undergo hydrolysis *in vivo* to form strongly fluorescent monomers. Therefore caution is warranted in relying on fluorescence detection (and other methods) with these particular sensitizers. In contrast, the sensitizers now under development are almost without exception based on a defined molecular structure. Thus fluorescence detection using these new sensitizers has the potential to become a quantitative diagnostic technique.

Table 1 shows that there is considerable variation in the fluorescence quantum yield (ϕ_F) of the sensitizers, ranging from near 0.5 for $AlPcS_4$ to <0.1 for DHE. Fluorescence excitation may be achieved using either the UV or red wavelength absorption bands, but in our experience (Barr et al 1988) excitation at longer wavelengths (e.g. 633 nm) is superior because the tissue autofluorescence is reduced.

A key question is whether fluorescence correlates primarily with *concentration or photoactivity for an individual sensitizer*. To understand this problem we must consider how the fluorescence quantum yield depends on the tissue microenvironment of the sensitizer and its physical state. For a dilute monomeric solution of a sensitizer such as $AlPcS_4$ in methanol, fluorescence correlates with the $AlPcS_4$ concentration. However, in heterogeneous biological systems the sensitizer is subject to complexation and fluorescence-quenching effects and may be partially aggregated. According to the excitation scheme shown earlier, which incorporates fluorescence quenching, the fluorescence quantum yield of monomeric sensitizer should correlate with the triplet state yield ($\phi_F \propto \phi_T$), assuming that the radiative properties of the singlet state and the rates of intersystem crossing and internal conversion are relatively insensitive to environmental effects. In other words, with allowance for fluorescence quenching, fluorescence intensity at a particular cellular site should correlate with the triplet yield and hence photoactivity, although different cellular sites have varying susceptibility to photosensitization. The other major factor is aggregation of the sensitizer, which in general considerably reduces the fluorescence efficiency and triplet-state yields; in the case of phthalocyanine aggregates (which are non-covalently bound) the fluorescence yield and photoactivity are negligible in comparison to those of the monomers (Harriman & Richoux 1980, Spikes & Bommer 1986). We therefore propose that fluorescence *in vivo* or *in vitro* should be taken primarily as a measure of photoactive monomer as opposed to total concentration. This could be experimentally verified *in vitro* by comparison of photosensitization action spectra and sensitizer absorption spectra. In summary, despite the varied experimental results using HpD/DHE, we would argue that fluorescence detection, when correctly applied, is of considerable relevance to PDT.

Photostability

Until recently, photostability was assumed to be desirable both for sensitizer storage and for optimum PDT efficacy. It is now apparent that the occurrence of photodegradation *in vivo* can be advantageous in terms of PDT selectivity. Photosensitizers are generally chosen to possess long-lived triplet states with high triplet-state and singlet oxygen quantum yields. Such dyes are particularly prone to photodegradation which can proceed by several mechanisms: firstly, singlet oxygen can react with the sensitizer, resulting in an oxidized photoproduct; and secondly, the excited photosensitizer triplet state can react with tissue components to form radical or ionic products.

Both DHE and AlPcS photodegrade during *in vivo* PDT (Mang et al 1987, Barr et al 1989a), as detected by loss of sensitizer fluorescence. For AlPcS this has been observed in PDT of normal rat colon, colonic rat tumour and mouse skin. In the experiments with normal colon, fluorescence was observed *in vivo* at 720–750 nm using 675 nm excitation, and, after treatment, *ex vivo* at 660–700 nm using 633 nm excitation. The *ex vivo* results for energy doses in the range of 100–200 J showed that more than 90% of the AlPcS was degraded in the central irradiation zone (diameter 4 mm) surrounding the point of contact with the optical fibre. At first these observations may appear surprising, given the exceptional photostability of AlPcS in aqueous solution (McCubbin & Phillips 1986) which results from AlPcS being highly resistant to attack by singlet oxygen, unlike many porphyrins. However, as discussed above, it is well known that phthalocyanines may undergo triplet state-activated reactions with other species, e.g. amino acids (Darwent et al 1982, Ferraudi et al 1988), and such type I processes should undoubtedly occur in tissue.

The extent of sensitizer photodegradation depends on the energy supplied and it has been proposed (Grossweiner 1986, Potter 1986) that for sufficiently low initial sensitizer concentrations the photosensitizer is degraded before tissue necrosis can take place. It can be shown that the initial threshold concentration C_0 required for phototoxicity is given by $C_0 > E^*k$, where E^* is the photodynamic threshold energy ($J\ cm^{-3}$) and k is the photodegradation rate constant. The value of k is determined by the photodegradation quantum yield, which need be only of the order of 10^{-3} (i.e. degradation after 10^3 absorption cycles) for extensive photodegradation to occur with typical PDT energy doses. Since normal tissue generally contains a lower amount of sensitizer than tumour, this provides a means of sparing adjacent normal tissue if it contains a sub-threshold concentration. The photodegradation mechanism has been proposed to account for improved selectivity of tumour treatment using DHE (Mang et al 1987) at lower doses. Likewise, Barr et al (1988a) have now shown with AlPcS that, for colonic rat tumour, treatment using a relatively low dose (0.5 mg/kg) enables selective necrosis of tumour without damage to normal tissue at the same energy dose (100 J), despite a mean AlPcS uptake ratio of only 2:1 between tumour and normal colon. Since AlPcS is observed to photodegrade during *in vivo* treatment of normal colon, these results are consistent with the benefits in selectivity of using lower doses of sensitizers which are proposed to result from photodegradation.

Acknowledgements

We are grateful for the close collaboration with The National Medical Laser Centre, University College, London and The Imperial Cancer Research Fund.

References

Barr H, Tralau CJ, MacRobert AJ, Morrison I, Phillips D, Bown SG 1988 Fluorescence photometric techniques for determination of microscopic tissue distribution of phthalocyanine photosensitisers for photodynamic therapy. Lasers Med Sci 3:81–86

Barr H, Tralau CJ, MacRobert AJ et al 1989a Selective necrosis in experimental colon cancer. Am J Gastroenterol, submitted

Barr H, Boulos PB, MacRobert AJ, Tralau CJ, Phillips D, Bown SG 1989b Comparison of lasers for photodynamic therapy with a phthalocyanine photosensitiser. Lasers Med Sci 4:7–12

Bonnett R, McGarvey DJ, Harriman A, Land EJ, Truscott TG, Winfield UJ 1988 Photophysical properties of m-tetraphenyl porphyrin and some m-tetra(hydroxyphenyl) porphyrins. Photochem Photobiol 48:271–276

Borland CF, McGarvey DJ, Morgan AR, Truscott TG 1988 Laser flash photolysis of porphyrins: novel potential photosensitisers of interest in photodynamic therapy. J Photochem Photobiol 2:427–435

Bown SG, Tralau CJ, Coleridge-Smith PD, Akdemir DTJ, Wieman TJ 1986 Photodynamic therapy with porphyrin and phthalocyanine sensitisation: quantitative studies in normal rat liver. Br J Cancer 54:43–52

Chan WS, Marshall JF, Svensen R, Phillips D, Hart IR 1987 Photosensitising activity of phthalocyanine dyes screened against tissue culture cells. Photochem Photobiol 45:757–761

Darwent JR, McCubbin I, Phillips D 1982 Excited singlet and triplet state electron transfer reactions of aluminium (III) sulphonated phthalocyanine. J Chem Soc Faraday Trans 2 78:347–357

Dougherty TJ 1987 Photosensitisers: therapy and detection of malignant tumours. Photochem Photobiol 45:879–889

Ferraudi G, Arguello GA, Ali H, van Lier JE 1988 Types I and II sensitised photo-oxidation of amino acids by phthalocyanines: a flash photochemical study. Photochem Photobiol 47:657–660

Firey PA, Jones TW, Jori G, Rodgers MAJ 1988 Photoexcitation of zinc phthalocyanine in mouse myeloma cells: the observation of triplet states but not of singlet oxygen. Photochem Photobiol 48:357–360

Grossweiner CI 1986 Optical dosimetry in photodynamic therapy. Lasers Surg Med 6:462–466

Harriman A, Richoux MC 1980 Attempted photoproduction of hydrogen using sulphonated phthalocyanines as chromophores for three component systems. J Chem Soc Faraday Trans 2 76:1618–1626

Keir WF, Land EJ, MacLennan AH, McGarvey DJ, Truscott TG 1987 Pulsed radiation studies of photodynamic sensitisers: the nature of DHE. Photochem Photobiol 46:587–589

Mang TS, Dougherty TJ, Potter WR, Boyle DG, Sommer S, Moan J 1987 Photobleaching of porphyrins used in photodynamic therapy and implications for therapy. Photochem Photobiol 45:501–506

McCubbin I, Phillips D 1986 The photophysics and photostability of zinc and aluminium sulphonated napthalocyanines. J Photochem 34:187–195

Phillips D 1989 Applications of lasers in photobiology and photochemistry. In: Radiation measurement in photobiology. Academic Press, p 103–133

Potter WR 1986 The theory of photodynamic therapy dosimetry: consequences of photodestruction of sensitiser. Proc International Society for Optical Engineering (SPIE) 712:124–129

Spikes JD, Bommer JC 1986 Zinc tetrasulphonated phthalocyanine as a photodynamic sensitiser for biomolecules. Int J Radiat Biol 50:41–45

Tralau CJ, Young AR, Walker NPJ et al 1989 Mouse skin photosensitivity with dihaematoporphyrin ether (DHE) and aluminium sulphonated phthalocyanine (AlPcS): a comparative study. Photochem Photobiol 49:305–312

Truscott TG, McLean AJ, Phillips AMR, Foulds WS 1988 Detection of haematoporphyrin derivative and haematoporphyrin excited states in cell environments. Cancer Lett 41:31–35

Valenzo DP 1987 Photomodification of biological membranes with emphasis on singlet oxygen mechanisms. Photochem Photobiol 46:147–160

Wilson BC, Patterson MS, Burns DM 1986 Effects of photosensitizer concentration in tissue on penetration depth of photoactivating light. Lasers Med Sci 1:235–244

DISCUSSION

Truscott: Dr MacRobert, you suggested that the S_1 excited state of the photosensitizer is not involved in producing cytotoxic species because it is too short lived. The lifetime is 10–15 ns, which is not extremely short. We also fail to detect singlet oxygen in cells because it's so short lived. Thus there is a contradiction here. The photosensitizer S_1 state may be short lived because it is very reactive and this may lead to the difficulty in detection. While I agree with you that photosensitizer triplet states and singlet oxygen are probably the most important agents in cytotoxicity, we should not rule out a role for the photosensitizer excited singlet state. In our laser studies of various sensitizers in cells, we see some evidence of species other than triplet states, possibly radicals.

MacRobert: We would agree that interaction of the excited singlet state with substrates to produce radical/ionic species cannot be completely ruled out. However, unambiguous experimental verification for such processes will be difficult to establish unless time-resolved detection techniques with (sub) nanosecond sensitivity are used.

Dougherty: John Parker (1987) claims to have identified singlet oxygen *in vivo*, but I know one group that has been unable to repeat that.

Truscott: I have not seen any report of reproduction of those results, but I have not tried the experiment myself.

Wilson: We found that singlet oxygen is not measurable *in vivo* with current techniques.

Dougherty: One incentive for us to find new sensitizers is that we might solve the problem of cutaneous photosensitivity. Dr MacRobert commented that sensitizers such as the phthalocyanines have an advantage in that they have very strong red absorptions relative to the Soret bands, whereas porphyrins have very strong Soret bands and low absorptions in the red. Because sunlight contains infrared as well as visible light, patients will be photosensitive whichever drug they receive, although they may not be as photosensitive if the drug does not have much additional absorption in the visible region. However, if the drug is in the skin, patients must stay out of the sun.

Bown: In a controlled study we investigated sensitization of nude mice with phthalocyanine and dihaematoporphyrin ether (DHE) (Tralau et al 1989). We used as the light source a solar simulator which emits a spectrum similar to that of sunlight. Using the same doses of sensitizing drugs and of activating light, we measured the minimum doses of simulated sunlight necessary for a skin reaction on mice. At all times after sensitization the light dose required to give a reaction with the phthalocyanine was about eight times greater than that to give a reaction with DHE.

Dougherty: That's true, but it's the wrong experiment. You must use 675 nm light, where the drug absorption peak is.

Bown: The patient would only be exposed to 675 nm light at the time of treatment. The problem is the exposure to sunlight at all times *after* treatment.

Dougherty: But what good is it to tell patients they can go outside for 15 minutes if they have been treated with a phthalocyanine and for 2.5 s if a porphyrin was used? You have to tell them to stay out of the sun for as long as they are photosensitive.

Bown: As Dr MacRobert described, another important factor in our search for new sensitizers is the ability to enhance selectivity of tumour damage by using low doses of the sensitizer.

Dougherty: That is also done with the porphyrins. We found that at a low dose of Photofrin II (0.5 mg/kg), which is still therapeutically useful, photosensitivity is markedly reduced, but not abolished. We still advise patients to stay out of the sun for a month and then to test themselves before exposure for long periods. The clinical problem is the same.

Bown: One must keep things in perspective. The light dose to produce a normal sunburn reaction in untreated mice was only double that required for an effect after sensitization with phthalocyanine (5 mg/kg). Phthalocyanines have not yet been used clinically, but patients should not be particularly sensitive to sunlight after PDT with these drugs.

Henderson: In my paper (Henderson & Bellnier, this volume) I present some data on the relationship between circulating drug and photosensitivity in the mouse. You have to be very careful what reaction you study. Dr Bown, were your comparisons done on the response in the foot or on the back skin?

Bown: We tried to do it on the whole animal, but the reaction with DHE was dramatic; the entire animal became swollen. Therefore, we irradiated a small spot on the back. We haven't looked at the foot pad response.

Jocham: The main aim of photodynamic therapy (PDT) is to cure the patient with minimal side-effects. I don't worry about photosensitization too much, but I like to be able to use a drug at low doses and with few side-effects. We need a drug that is selective for tumours and effective at low doses. Reducing the dosage and achieving minimal effects within normal tissue is only acceptable if strong effects are retained within the tumour. Therapeutic action only within the range of 2 mm in the tumour is not acceptable.

Kaye: Whether photosensitivity is an important issue depends what you are treating. In treatment of a potentially lethal tumour it doesn't matter how photosensitive patients become; the priority is to ensure maximum tumour kill. But when treating a less serious problem you may need to worry about skin photosensitivity.

Brown: It's clear that if a patient is to be cured, then 2–3 months out of sunlight in return for several years of life is not very serious. But if, as in many cases, the treatment simply provides palliation in return for survival for perhaps six months or a year, three months out of sunlight is quite a significant factor. That's an issue we must consider seriously in deciding whether to treat patients with PDT.

Jori: In the treatment of skin tumours, one needs to selectively kill the tumour while sparing the peritumoral tissue. Using both the back skin and the foot pad tests, we also find that if we irradiate experimental animals with white light there is much less photosensitivity with phthalocyanines than with porphyrins.

The Soret bands for phthalocyanines are at 350 nm. This is very important. Particular wavelengths in sunlight do not have the same intensity throughout the day. At midday the most intense sunlight is in the blue region, where porphyrins absorb strongly and phthalocyanines do not absorb as much. Red light is intense only in the early morning and at sunset in normal atmospheric conditions. Thus we think phthalocyanines carry less risk of skin photosensitivity than porphyrins.

Moan: It has long been known that photobleaching of Photofrin II takes place under clinical conditions. This is not a disadvantage but adds another dimension to photodynamic therapy: if you have a drug that is several times more concentrated in tumours than in normal tissues, you may be able to cure a tumour at any depth if you can get a high enough light dose to it. You can choose a low dose of drug which, because of bleaching, will not kill the normal cells however much light is used. The tumour contains more drug and will be destroyed before the drug is bleached.

Dougherty: The practical limitation to that is the penetration of the wavelength used in the tissue treated. One wants to use light that penetrates deeper but with photosensitizers that still take advantage of the photobleaching effect.

Moan: The only limitation is the time that you can keep a patient under the light.

Dougherty: What were the relative efficiencies of photosensitization in rat liver for the porphyrin (630 nm) and phthalocyanine (675 nm)?

MacRobert: The production of a 4 mm diameter lesion using the same dose of each sensitizer (5 mg/kg) required ten times less energy for the phthalocyanine than HpD (Bown et al 1986).

Dougherty: The liver is perhaps the wrong model to use because its optical properties are dominated by absorption, whereas those of most tissues are dominated by scattering. The scattering differential between 630 and 675 nm is almost insignificant and that will be the dominant factor in most tissues.

Wilson: Yes; it may be misleading to generally expect that improvement in penetration in going from porphyrins to phthalocyanines.

Dougherty: We are all looking for new sensitizers that have deeper therapeutic effects, and we are working at activating wavelengths of around 800 nm where there is optimum tissue penetration. If we find a sensitizer with twice the effective penetration, which is all we can expect, what new clinical applications will be possible?

Kaye: Provided you retain the tumour selectivity, treatment of brain tumours would be enhanced by a substantial increase in the depth of tumour kill.

Jocham: The answer to this question depends on the tumour and on the organ. PDT may cure or at least provide palliation for lung tumours if the penetration of treatment is sufficient. For example, this may be preferable to using a YAG laser in treating obstructing lung tumours, because there is little risk of inducing bleeding in the area behind the tumour. But PDT offers less advantage in treating infiltrating bladder tumours, such as a TIII tumour within the bladder tissue that is bound to the deep muscle layer, where there is a high risk of metastasis. In this case it is preferable to remove tumour material using an electric loop.

Dougherty: For lung tumours the current PDT technique works as well as any possibly could. Greater penetration is not necessary because the treatment already kills everything right to the bronchial wall.

Jocham: That depends on whether you combine PDT with some other type of treatment.

Wilson: It might be possible to treat head and neck tumours and also prostatic tumours. The usefulness of PDT for treating larger volumes may depend on the development of some sort of fractionation scheme. One problem in the treatment of large volumes of internal organs is the clearance of the dead tissue which results from the rapid necrosis produced by single fraction light treatments.

Bonnett: Surely large tumours will not be treated by this method? If they are operable, large solid tumours will presumably be excised. Subsequent clean-up of the wound by phototherapy is a possibility.

Hönigsmann: If a drug has greater effective penetration, it will also have increased penetration into skin, which might lead to more side-effects.

Morgan: We seem to be assuming that new photosensitizers will remain in skin for three months. There are already sensitizers that are completely cleared from skin 48 hours after administration. So we are worrying about effects that may apply only to Photofrin II.

Dougherty: Phthalocyanines are retained for a long time.

Morgan: I don't think they will be approved for clinical use.

Lim: The alternative to developing a photosensitizer with a very short half-life is to develop a sunscreen which blocks out the visible to infrared spectrum. Most sunscreens work well against UVB and some work against UVA below 400 nm, but there is no incentive to develop a sunscreen that blocks out the rest of the spectrum because most dermatological problems result from wavelengths below 400 nm.

Dougherty: Would such sunscreens be cosmetically acceptable?

Lim: The only available ones that could block the rest of the spectrum belong to the zinc oxide family and are not cosmetically acceptable.

Hönigsmann: Have you tried to give β-carotene after PDT?

Dougherty: That's probably not practical. Although β-carotene works, it takes a long time to reach the necessary level in skin. Bolus injection at high doses doesn't work, because it is excreted too quickly.

Lim: In porphyria, β-carotene is thought to work not as a sunscreen but as a quencher of oxygen radicals.

Dougherty: Yes, but that would also work here. We demonstrated a protective effect in mice many years ago. The problem is that it takes about 30 days to build up the required level in the skin. By that time the photosensitivity has passed anyway.

Truscott: Canthaxanthin/β-carotene mixtures are used for porphyria patients. There may be other carotenoids which accumulate more rapidly in skin.

Dougherty: In theory, that's true. But it would be better to solve this problem without putting more drugs in the patient.

References

Bown SG, Tralau CJ, Coleridge-Smith PD, Akdemir D, Wieman TJ 1986 Photodynamic therapy with porphyrin and phthalocyanine sensitisation: quantitative studies in normal rat liver. Br J Cancer 54:43–52

Henderson BW, Bellnier DA 1989 Tissue localization of photosensitizers and the mechanism of photodynamic tissue destruction. In: Photosensitizing compounds: their chemistry, biology and clinical use. Wiley, Chichester (Ciba Found Symp 146) p 112–130

Parker J 1987 Optical monitoring of singlet oxygen generation during photodynamic treatment of tumours. IEEE (Inst Electr Electron Eng) Circ Dev, p 10–21

Tralau CJ, Young AR, Walker NPJW et al 1989 Mouse skin photosensitivity with dihaematoporphyrin ether (DHE) and aluminium sulphonated phthalocyanine (AlSPc): a comparative study. Photochem Photobiol 49:305–312

The chemistry, photophysics and photosensitizing properties of phthalocyanines

Johan E. van Lier* and John D. Spikes†

*MRC Group in the Radiation Sciences, Faculty of Medicine, University of Sherbrooke, Sherbrooke, Quebec, Canada J1H 5N4 and †Department of Biology, University of Utah, Salt Lake City, UT 84112, USA

Abstract. Phthalocyanines (Pcs) and naphthalocyanines (Ncs) are being extensively studied as photosensitizers for photodynamic therapy (PDT) of cancer. They strongly absorb clinically useful red light, with maxima around 670 nm and 770 nm respectively. Chelated with appropriate diamagnetic metal ions, they exhibit high triplet yields and long triplet lifetimes. Energy transfer from the triplet dye to ground-state oxygen to yield singlet oxygen appears to be the main photosensitizing pathway in biological systems. Underivatized Pcs and Ncs can be incorporated in liposomes for *in vivo* administration. Sulphonation renders the dyes water soluble but also enhances dimerization to yield photochemically inactive aggregates. Tumour retention and cell membrane penetration of the dyes are strongly affected by the polarity of the macrocycle as well as the nature of the central metal ion and axial ligands. Among the sulphonated dyes, amphiphilic mono- and disulphonated derivatives exhibit particularly good cell membrane-penetrating properties, although the more highly sulphonated dyes show better tumour retention *in vivo*. At least *in vitro*, Pc dyes are more photoactive than the corresponding Nc dyes, which probably reflects the lower photostability of the latter.

1989 Photosensitizing Compounds: their Chemistry, Biology and Clinical Use. Wiley, Chichester (Ciba Foundation Symposium 146) p 17–32

The sensitizer preparation currently used in clinical trials of photodynamic therapy (PDT) of cancer consists of a mixture of haematoporphyrin derivatives (HpD) (Dougherty 1987). Haematoporphyrins, however, absorb only weakly above 600 nm where light exhibits the deepest penetration into tissues. Furthermore, the uptake by the tumour of the active component in HpD is only marginally higher than that by most normal tissues. Thus it has been evident for some time that the effectiveness of PDT could be enhanced by the use of photosensitizers that absorb more strongly towards the red end of the light spectrum and are more selectively retained by neoplastic tissues. For these reasons, several new classes of photosensitizers for PDT have been suggested over the past few years; among them, phthalocyanines (Pcs) and the closely related naphthalocyanines (Ncs) have received increasing attention (for reviews see Spikes 1986, Ben-Hur 1987, van Lier et al 1988). There is an extensive literature on phthalocyanines relating to their numerous commercial applications as catalysts of redox reactions, as semiconductors, as light absorbers in photoelectric cells as well as pigments in inks and paints (Moser & Thomas 1983).

Chemistry

Phthalocyanines are azaporphyrin derivatives which mimic naturally occurring porphyrins in many aspects. Their central macrocycle consists of a cyclic tetrapyrrole unit, but, in contrast to porphyrins, the pyrrole subunits in Pc and Nc are linked by nitrogen atoms rather than methine bridges (Fig. 1). Conjugation of the macrocycle is extended by one or two benzo rings (Pc or Nc respectively) on the four pyrrole subunits, resulting in strong absorption bands in the clinically useful red region of the spectrum. They form stable complexes with many metal ions and several metallo complexes can be prepared directly by chelating metal cations with the four central pyrrole nitrogens of the macrocycle. Complexes with metal ions or oxides whose radii are too large for direct insertion into the Pc or Nc central pocket can be prepared via ring closure of precursors of the macrocycle in the presence of the appropriate metal salt to give out-of-plane and sandwich structures (Weber & Busch 1965). Pc and Nc and their metal complexes are insoluble in water. However, Zn-Pc can

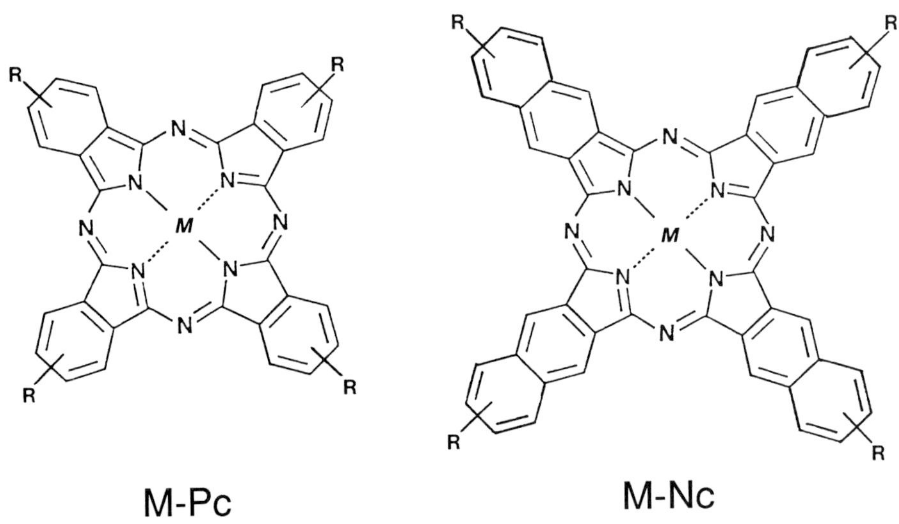

FIG. 1. Structures of substituted metallo-phthalocyanine (M-Pc) and naphthalocyanine (M-Nc). R = H, SO_3^-, NH_4^+, NO_2, OH, COOH, etc.

be incorporated into the membranes of liposomes and thus be made compatible with physiological media (Valduga et al 1988).

Most biological studies on Pcs related to PDT have been conducted with water-soluble sulphonated metallo-phthalocyanines (M-PcS). These can be obtained by sulphonation of the appropriate M-Pc in oleum (Linstead & Weiss 1950) or by the condensation of mixtures of sulphonated precursors in the presence of the selected metal salt (Weber & Busch 1965). Although both condensation and direct sulphonation procedures lead to mixtures of mono- to tetrasulphonated products, reaction conditions can be selected such that good yields of rather clean products are obtained (Ali et al 1988). The separation of the many possible isomers among the substituted dyes requires extensive chromatography on reverse-phase columns using solvent gradients. The average degree of sulphonation of these Pc and Nc dyes can be rapidly determined by oxidative degradation of the macrocycle followed by quantitative chromatographic analysis of the resulting substituted pyrrole fragments (Ali et al 1988). Combined with chromatographic analysis, this procedure provides a simple way to characterize batches of mixed sulphonated dye preparations for use in biological experiments. M-Pcs substituted with other physiologically attractive functional groups, such as amino, carboxyl, nitro, hydroxy and neopentoxy groups, have been evaluated for PDT, while the photodynamic properties of positively charged amino-Pc and pyridino-Pc analogues are currently under investigation (see reviews cited above and Rosenthal et al 1987). Naphthalocyanines have more recently been added to the growing list of potential second-generation photosensitizers for PDT. Initial studies have centred mainly on their photophysics (McCubbin & Phillips 1986, Firey et al 1988a). Compounds in this class include lipophilic substituted silicon-Ncs and the water-soluble sulphonated Zn- and Al-Nc, prepared in a manner similar to the analogous M-PcS.

Photophysical properties

Monomeric metallo-Pcs in aqueous media have characteristic absorption spectra: a Soret band at approximately 350 nm; a small band around 600 nm; and a narrow, very strong absorption peak (Q band) around 670 nm (molar extinction coefficient (ϵ) in the range of 10^5 $M^{-1}cm^{-1}$) (Darwent et al 1982). Absorption maxima of the naphthalocyanine Q bands are red-shifted a further 100 nm (λ_{max} 740–780 nm), with similar high extinction coefficients (Fig. 2). In addition to being red-shifted, Pc and Nc light absorption is almost two orders of magnitude stronger than the highest Q band absorption of haematoporphyrin derivatives ($\lambda_{max} = 630$ nm, $\epsilon \sim 10^3$ $M^{-1}cm^{-1}$). Absorption maxima of water-soluble M-PcSs may vary slightly with the pH. Such variations are attributed to changes in the axial ligands coordinated to the metal centre rather than protonation of macrocycle sites (Ferraudi et al 1988). For example, shifts in

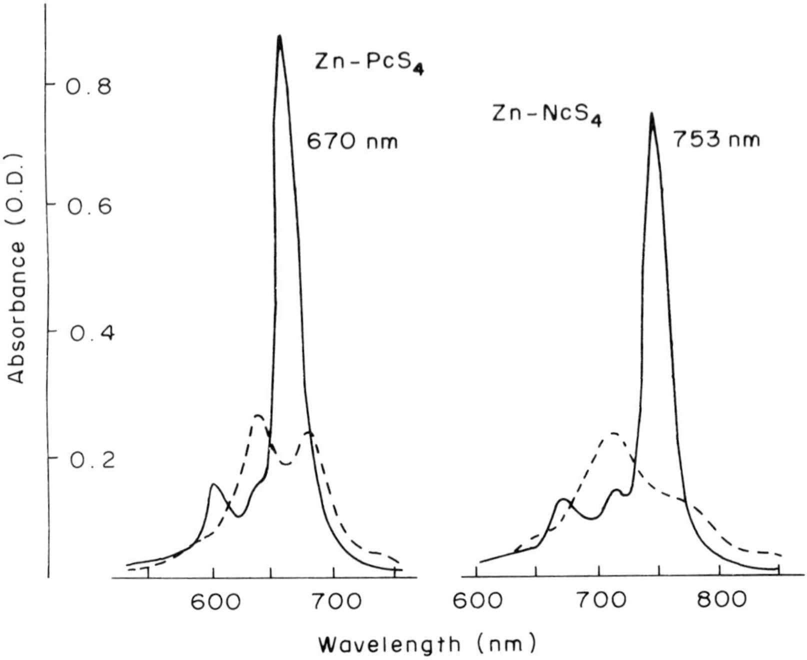

FIG. 2. Absorption spectra of 4 μM tetrasulphonated zinc phthalocyanine ($ZnPcS_4$) and naphthalocyanine ($ZnNcS_4$) for the aggregated dyes in water (---) and the monomeric dyes in 95% methanol (—).

the Soret and Q bands (~10 nm) of the tetrasulphonated $AlPcS_4$ are observed as a function of pH in a manner consistent with the existence of two acid–base equilibria for the deprotonation of axially coordinated water (Equation 1).

$$AlPcS_4(OH_2)_2^{3-} \leftrightharpoons AlPcS_4(OH_2)OH^{4-} \leftrightharpoons AlPcS_4(OH)_2^{5-} \quad (1)$$

In biological fluids, most M-PcSs dimerize and aggregate, resulting in loss of photochemical activity. Spectroscopically, aggregation of Pc and Nc dyes can be recognized by a shift to shorter wavelengths of the Soret band and a broadening, together with a 30–50 nm blue-shift, of the Q band (Fig. 2) (Bernauer & Fallab 1961). The tendency of phthalocyanines to aggregate in aqueous solutions is strongly influenced by the type and degree of substitution. Non-substituted Zn-Pc incorporated in liposomes is photochemically active (Valduga et al 1988) and exhibits an absorption spectrum similar to that of monomeric $ZnPcS_4$ in water (Spikes & Bommer 1986). Dilute aqueous solutions of tetrasulphonated $GaPcS_4$ and $AlPcS_4$ show little aggregation. Decreasing the number of sulphonate substituents results in increased aggregation, with K_d values reaching $4.5 \times 10^6\ M^{-1}$ (50% MeOH in water) for the disulphonated $GaPcS_2$. The capacity to generate singlet oxygen decreases accordingly. However, the degree of substitution does not appear to affect the photochemical properties of the M-Pc *per se*, but rather the extent of aggregation determines the actual photochemical activities of the dyes (Wagner et al 1987). Disaggregation can be accomplished by added detergents or organic solvents. Monomeric Pcs fluoresce strongly, with a single peak in the 680–690 nm range, whereas aggregated Pcs do not fluoresce appreciably. The fluorescence lifetime of $AlPcS_4$ in aqueous medium is approximately 6 ns (Darwent et al 1982).

The photophysical properties of the Pc dyes are strongly influenced by the presence and nature of the central metal ion. This is reflected in the large variations in yield (Φ_T) and lifetimes (τ_T) of the photoexcited triplet states of M-Pc. Complexation of Pc with transition metals gives dyes with short triplet lifetimes (in the ns range), whereas closed d shell diamagnetic ions, such as Zn^{2+}, Al^{3+} and Ga^{3+}, give Pc complexes with both high triplet yields ($\Phi_T > 0.4$) and long lifetimes ($ZnPcS_4$, $\tau_T = 240\ \mu s$; $AlPcS_4$, $\tau_T = 400\ \mu s$) (Darwent et al 1982). Consequently the latter complexes can be expected to exhibit the strongest photochemical and photodynamic activities. Triplet M-Pc can be quenched by either an electron transfer reaction (type I) to yield an anion $Pc^{\overline{\cdot}}$ radical and a positively charged substrate radical ($S^{\cdot +}$) (Equation 2), or by an energy transfer process (type II). The type II process usually involves conversion of ground-state triplet oxygen ($^3O_2$) into $^1\Delta_g$ singlet oxygen ($^1O_2$), as depicted in Equation 3.

$$(T_1)Pc + S \xrightarrow{\text{(type I)}} Pc^{\overline{\cdot}} + S^{\cdot +} \xrightarrow{O_2} SOOH \qquad (2)$$

$$(T_1)Pc + O_2 \xrightarrow{\text{(type II)}} Pc + {}^1O_2 \xrightarrow{S} SOOH \qquad (3)$$

The triplet energies of Pcs vary from 110–126 kJ/mol, which is sufficient to generate $^1O_2$ (94.5 kJ/mol), and high quantum yields, in the range of 0.3–0.5, have been reported for $^1O_2$ generation by M-Pc (Rosenthal et al 1986, Wagner et al 1987). Singlet oxygen, which in water has a lifetime of 4.2 μs, readily reacts with a variety of unsaturated biomolecules to give intermediate hydroperoxy products (Equation 3). Likewise, type I processes will result in hydroperoxide formation through the reaction of substrate-centred cation radicals with ground-state oxygen (Equation 2). Thus, both type I and II pathways lead to oxidative degradation of cellular components, the actual type of damage depending on the localization of the dye and oxygen availability. Although electron transfer quenching of (T_1)M-Pc by organic molecules in the absence of oxygen has been observed, rate constants for (T_1)M-Pc quenching by ground-state $^3O_2$ are generally several orders of magnitude higher (Ohno et al 1983, Spikes & Bommer 1986, Ferraudi et al 1988). In accord with this, both cholesterol and tryptophan give exclusively the characteristic $^1O_2$ oxidation products during photosensitized oxidation by water-soluble M-PcS (Langlois et al 1986). It thus appears that only close association of M-Pc with specific amino acid residues, such as tyrosine, combined with low oxygen concentrations, would lead to conditions where a type I pathway could represent a significant contribution to protein damage under physiological conditions. Nevertheless, Firey et al (1988b) failed to detect $^1O_2$ infrared luminescence with Zn-Pc in cellular media following flash illumination while observing Zn-Pc triplet states. They point out, however, that the most logical rational for their failure to observe $^1O_2$ luminescence, and the earlier observed lack of an *in vitro* D_2O effect (Ben-Hur & Rosenthal 1985), is that singlet oxygen must be extremely rapidly quenched in the cellular milieu. On balance, $^1O_2$ formation is most probably the principal mechanism of photosensitized biological damage by M-Pc derivatives.

The photophysical properties of sulphonated metallo-Nc resemble those of the corresponding Pc, and, like the latter, they show a pronounced tendency to aggregate in aqueous solution (McCubbin & Phillips 1986). However, aggregated M-NcS fluoresces in aqueous solution, a phenomenon not observed with the corresponding M-PcS. The triplet lifetimes of naphthalocyanines are shorter than those of the corresponding phthalocyanines, but metal substitution within the Nc macrocycle is likely to result in the same photophysical effects as within the Pcs. The greater conjugation of the Nc may stabilize the $Nc^{\cdot +}$ cation, thus facilitating photoinduced electron transfer reactions. The increase in ligand size renders Ncs three orders of magnitude more sensitive to photo-oxidation than Pcs (McCubbin & Phillips 1986). In addition to chemical quenching, energy transfer from triplet Si-Nc to ground-state $^3O_2$ was shown to produce $^1O_2$ in a reversible reaction (Firey et al 1988a). Fluorescence at 1330 nm is attributed to the triplet to ground state decay of the photoexcited Si-Nc, which corresponds to an energy gap of 90.3 kJ/mol. The $^1O_2$ to $^3O_2$ transition is at 94.5 kJ/mol (1269 nm), and accordingly the energy transfer between (T_1)Si-Nc and $^3O_2$ is endoergic. Despite this, the efficiency of this conversion in a homogeneous solution is nearly 100% because the equilibrium state is rapidly drained, mainly via the decay of the relatively short-lived $^1O_2$ (Firey et al 1988a).

Photodynamic activities

Numerous accounts of both *in vitro* and *in vivo* photodynamic activities of various Pc derivatives have appeared over the past few years (see reviews cited in the introduction). Several M-Pcs including water-soluble sulphonated analogues, exhibit *in vitro* photoactivities superior to that of HpD. The relative phototoxicity varies substantially between different Pc derivatives and some structure–activity relationships have been proposed. The effect of the degree of sulphonation of M-Pcs on their photodynamic properties has been studied in detail. Although the less sulphonated dyes exhibit reduced photochemical activity because of their higher tendency to form photochemically inactive aggregates, they show an increased phototoxicity *in vitro* toward V-79 Chinese hamster cells; the mono- and disulphonated derivatives of Ga- and AlPcS are the most active, whereas the tri- and tetrasulphonated dyes are inefficient sensitizers of cell killing (Brasseur et al 1987, Paquette et al 1988). The same pattern of photoactivity versus degree of sulphonation is observed for ZnPcS, except that addition of the monosulphonated dye to the disulphonated preparation results in decreased photoactivity, probably because of the poor solubility of the monosulphonated dye (Brasseur et al 1988). Comparison of the photocytotoxicity towards V-79 cells of disulphonated Zn-, Al- and GaPcS with that of HpD gives the relative efficiencies 36:2:0.5:1, suggesting $ZnPcS_2$ as the most promising candidate in this series for PDT. Thus, in addition to the type and degree of substitution on the Pc macrocycle, the nature of the central metal ion strongly influences the overall activity of the dye.

The importance of cell-penetrating properties of the M-PcS dyes in exerting biological activity has been studied by fluorescence microscopy of V-79 cells. Incubation with variously sulphonated AlPcS revealed that the highly photocytotoxic, disulphonated derivative gave uniform fluorescence in the cell cytoplasm whereas cells incubated with the photoinactive tri- and tetrasulphonated derivatives showed very little dye uptake (Paquette et al 1988). Studies on the four constitutional isomers of the disulphonated $GaPcS_2$ showed that the most hydrophilic isomer, with sulphonate groups at opposite positions of the Pc molecule, is completely void of photocytotoxicity, whereas the most hydrophobic isomer, with sulphonate substituents at adjacent benzene rings, is extremely efficient as a photosensitizer of V-79 cell killing (Brasseur et al 1987). This suggests that only amphiphilic isomers, with sulphonate substituents located on one side of the Pc macrocycle, are capable of crossing the cell membrane and subsequently exerting phototoxicity.

Whereas *in vitro* photosensitization of cell killing is mainly associated with the photochemical and cell-penetrating properties of the dye, *in vivo* PDT activity appears to be governed by many more parameters. The outcome with most sensitizers is destruction of the vascularity rather than direct tumour cell kill, and is probably not related to the tumour cell-penetrating properties of the dye. However, tumour retention of the dye *in vivo*, like cell uptake *in vitro*, is a basic requirement for successful PDT. Several studies of the pharmacokinetics and tumour uptake of different Pcs have been reported. Zn-Pc incorporated into unilamellar liposomes is selectively retained in the tumours after i.p. injection in BALB/c mice bearing a transplanted MS-2 fibrosarcoma (Reddi et al 1987). The non-substituted Zn-Pc dye is transported by serum lipoproteins and cleared from the serum via the bile–gut pathway. Zn-Pc is rapidly cleared from most tissues after a maximum uptake within 3 h, whereas the tumour concentration reaches a maximum only after 24–48 h. At this time the ratio of the tumour and muscle concentrations is about 7.5 and exposure to red light results in extensive tumour necrosis; both the vascular system and neoplastic cells exhibit signs of severe photodamage (Milanesi et al 1987). Water-soluble sulphonated Pcs are also retained to some extent by experimental tumours in rodents, and studies with ^{14}C-labelled GaPcS show that the more highly sulphonated dyes reach the greatest tumour concentrations (J. Rousseau et al, unpublished work 1988). For the trisulphonated dye, tumour-to-tissue ratios of nearly 20 are observed 48 h after the injection. However, such highly sulphonated dyes are unlikely to penetrate tumour cells: the observed effects of PDT presumably result exclusively from vascular photodamage. Mixtures of tri- and tetra-sulphonated AlPcS induce extensive haemorrhagic necrosis 24 h after PDT in a transplantable bladder tumour in the rat, an effect that coincides with a reduction of the tumour blood flow (Selman et al 1986). Using a similar AlPcS preparation in a transplantable fibrosarcoma in rats, Tralau et al (1987) found that the extent of tumour necrosis is predictable and directly related to the *in situ* light dose and dye concentration. Thus, the degree of sulphonation of the M-Pc not only affects the biodistribution pattern but, as already mentioned, also strongly influences the cell-penetrating properties of the dyes. *In vitro*, variations in the number of Pc substituents leads to large differences in photocytotoxicity, but the effect *in vivo* is far less pronounced (Brasseur et al 1988). Water-soluble, differently sulphonated Al-, Ga- and Zn-phthalocyanines have been tested for their relative effectiveness as photosensitizers for the PDT of EMT-6 mammary tumours in BALB/c mice. Although the tumour response follows a similar structure–activity relationship to that observed under *in vitro* conditions, extreme variations are not observed (Brasseur et al 1988). Disulphonated ZnPcS is the most active photosensitizer in this series, and the symmetric tetrasulphonated ZnPcS, prepared by the condensation method, shows no activity (at a dose of <12 mg/kg). The corresponding disulphonated $GaPcS_2$ and $AlPcS_2$ show activities comparable to HpD at similar dose levels. Thus, based on the elevated *in vitro* activities, the desired *in vivo* PDT effect requires the administration of higher than expected doses of Pc dyes. We shall require a more detailed understanding of the structure–activity relationships of substituted Pcs and Ncs and their interaction with cellular components to fully exploit their potential in photodynamic therapy of cancer.

The biological activities of the naphthalocyanines are being studied in several laboratories, but pertinent data on their photodynamic properties are still scarce. Further studies on the potential of Nc dyes as photosensitizers for PDT of cancer are made particularly attractive by the availability of inexpensive diode lasers which deliver light at the IR absorption maxima.

Acknowledgements

The authors are grateful to the Medical Research Council of Canada, the National Cancer Institute of Canada, the American Cancer Society (grant PDT-259), the DOD/SDIO Medical Free Electron Laser Program, and the Utah Laser Institute for financial support.

References

Ali H, Langlois R, Wagner JR, Brasseur N, Paquette B, van Lier JE 1988 Biological activities of phthalocyanines. X. Syntheses and analyses of sulfonated phthalocyanines. Photochem Photobiol 47:713–717

Ben-Hur E, Rosenthal I 1985 Photosensitized inactivation of Chinese hamster cells by phthalocyanines. Photochem Photobiol 42:129–133

Ben-Hur E 1987 Photochemistry and photobiology of phthalocyanines: new sensitizers for photodynamic therapy of cancer. In: Favre A et al (eds) From photophysics to photobiology. Elsevier Science Publishers, New York, p 407–420

Bernauer K, Fallab S 1961 Phthalocyanine in wasseriger Losung I. Helv Chim Acta 44:1287–1292

Brasseur N, Ali H, Langlois R, van Lier JE 1987 Biological activities of phthalocyanines. VII. Photoinactivation of V-79 Chinese hamster cells by selectively sulfonated gallium phthalocyanines. Photochem Photobiol 46:739–744

Brasseur N, Ali H, Langlois R, van Lier JE 1988 Biological activities of phthalocyanines. IX. Photosensitization of V-79 Chinese hamster cells and EMT-6 mouse mammary tumor by selectively sulfonated zinc phthalocyanines. Photochem Photobiol 47:705–711

Darwent JR, Douglas P, Harriman A, Porter G, Richoux M-C 1982 Metal phthalocyanines and porphyrins as photosensitizers for reduction of water to hydrogen. Coord Chem Rev 44:83–126

Dougherty TJ 1987 Photosensitizers: therapy and detection of malignant tumors. Photochem Photobiol 45:879–889

Ferraudi G, Arguello GA, Ali H, van Lier JE 1988 Types I and II sensitized photooxidation of aminoacid by phthalocyanines: a flash photochemical study. Photochem Photobiol 47:657–660

Firey PA, Ford WE, Sounik JR, Kenney ME, Rodgers MAJ 1988a Silicon naphthalocyanine triplet state and oxygen: a reversible energy-transfer reaction. J Am Chem Soc 110:7626–7630

Firey PA, Jones TW, Jori G, Rodgers MAJ 1988b Photoexcitation of zinc phthalocyanine in mouse myeloma cells: the observation of triplet states but not of singlet oxygen. Photochem Photobiol 48:357–360

Langlois R, Ali H, Brasseur N, Wagner JR, van Lier JE 1986 Biological activities of phthalocyanines. IV. Type II sensitized photooxidation of L-tryptophan and cholesterol by sulfonated metal phthalocyanines. Photochem Photobiol 44:117–123

Linstead RP, Weiss FT 1950 Phthalocyanines and related compounds. Part XX. Further investigations on tetrabenzporphin and allied substances. J Chem Soc 2975–2987

McCubbin I, Phillips D 1986 The photophysics and photostability of zinc(II) and aluminium(III) sulphonated naphthalocyanines. J Photochem 34:187–195

Milanesi C, Biolo R, Reddi E, Jori G 1987 Ultrastructural studies on the mechanism of the photodynamic therapy of tumors. Photochem Photobiol 46:675–681

Moser FH, Thomas AL 1983 The phthalocyanines. Vol I and II. CRC Press, Boca Raton, Florida

Ohno T, Kato S, Yamada A, Tanno T 1983 Electron transfer reactions of the photoexcited triplet state of chloroaluminium phthalocyanine with aromatic amines, benzoquinones, and coordination compounds of iron(II) and iron(III). J Phys Chem 87:775–781

Paquette B, Ali H, Langlois R, van Lier JE 1988 Biological activities of phthalocyanines. VIII. Cellular distribution in V-79 Chinese hamster cells and phototoxicity of selectively sulfonated aluminium phthalocyanines. Photochem Photobiol 47:215–220
Reddi E, LoCastro G, Biolo R, Jori G 1987 Pharmacokinetic studies with zinc(II)-phthalocyanine in tumour-bearing mice. Br J Cancer 56:597–600
Rosenthal I, Krishna CM, Riesz P, Ben-Hur E 1986 The role of molecular oxygen in the photodynamic effect of phthalocyanines. Radiat Res 107:136–142
Rosenthal I, Ben-Hur E, Greenberg S, Concepcion-Lam A, Drew DM, Leznoff CC 1987 The effect of substituents on phthalocyanine phototoxicity. Photochem Photobiol 46:959–963
Selman SH, Kreimer-Birnbaum M, Chaudhuri K et al 1986 Photodynamic treatment of transplantable bladder tumors in rodents after pretreatment with chloroaluminium tetrasulfophthalocyanine. J Urol 136:141–145
Spikes JD 1986 Phthalocyanines as photosensitizers in biological systems and for the photodynamic therapy of tumors. Photochem Photobiol 43:691–699
Spikes JD, Bommer JC 1986 Zinc tetrasulfophthalocyanine as a photodynamic sensitizer for biomolecules. Int J Radiat Biol 50:41–45
Tralau CH, MacRobert AJ, Coleridge-Smith PD, Barr H, Bown SG 1987 Photodynamic therapy with phthalocyanine sensitization: quantitative studies in a transplantable rat fibrosarcoma. Br J Cancer 55:389–395
Valduga G, Nonell S, Reddi E, Jori G, Braslavsky SE 1988 The production of singlet molecular oxygen by zinc(II) phthalocyanine in ethanol and unilamellar vesicles. Chemical quenching and phosphorescence studies. Photochem Photobiol 48:1–5
van Lier JE, Brasseur N, Paquette B et al 1988 Phthalocyanines as sensitizers for photodynamic therapy of cancer. In: Moreno G et al (eds) Photosensitization: molecular, cellular and medical aspects. Springer-Verlag, Berlin, p 435–444
Wagner JR, Ali H, Langlois R, Brasseur N, van Lier JE 1987 Biological activities of phthalocyanines. VI. Photooxidation of L-tryptophan by selectively sulfonated gallium phthalocyanine: singlet oxygen yields and effect of aggregation. Photochem Photobiol 45:587–594
Weber JH, Busch DH 1965 Complexes derived from strong field ligands. XIX. Magnetic properties of transition metal derivatives of 4,4′,4″,4‴-tetrasulfophthalocyanine. Inorg Chem 4:469–471

DISCUSSION

Dougherty: Phthalocyanines (Pcs) have much stronger absorptions in the red than porphyrins and have similar singlet oxygens yields. However, in *in vivo* experiments they are inefficient; the uptake in tumours is high, and yet tumour destruction requires the same amount of light as with porphyrins. Why can't we take advantage of the greater absorption?

van Lier: We hope that autoradiography with the ^{14}C-labelled sulphonated phthalocyanines will answer that. Efficiency of tumour kill must be related not only to the tissue distribution of the photosensitizer, but even more so to the distribution within the tumour. Apparently many of these substituted Pcs localize in areas where they cannot do as much damage as anticipated.

Dougherty: Is this related to the degree of sulphonation of the dyes?

van Lier: Yes; the more they are sulphonated the better they localize in the capsule around the tumour that is rich in freshly laid down collagens.

Jori: Have you used liposomal injections?

van Lier: No. The monosulphonated metallo-Pcs, which are efficient generators of singlet oxygen and have excellent cell-penetrating properties, would be the ones to compare in liposomes with dyes that have no functional groups at all.

Dougherty: Does using liposomes change the distribution of the drug?

Jori: Yes; we have compared the phototherapeutic efficiency of a mixture of water-soluble mono- and disulphonated phthalocyanines with that of a non-sulphonated zinc phthalocyanine, which can be injected only when incorporated into the phospholipid bilayer of liposomes. The effects on the tumour tissues were quite different, suggesting different distributions of the drug within the tumour. Electron microscopy showed that the sulphonated mixture caused mainly vascular damage of irradiated tumour tissues, whereas liposome-delivered zinc phthalocyanine caused mainly direct damage to neoplastic cells.

Dougherty: It would be interesting to compare delivery of Dr van Lier's mixture of sulphonated phthalocyanines in aqueous solution and in liposomes.

van Lier: Yes. That would be a good way to study the effect of drug formulation on tumour retention and photodynamic tissue damage.

Henderson: Various factors may limit the photodynamic potential of phthalocyanines. We also have been surprised that they are not better agents for tumour destruction. Contrary to what one might expect, these drugs limit light transmission. Forty percent of 680 nm incident light is transmitted through a 3 mm RIF tumour plus skin in the absence of phthalocyanine. With 10 mg/kg of Ga-Pc there was about 23% transmission. In control mice, 37% of incident light is transmitted at 630 nm and almost as much is transmitted when 10 mg/kg Photofrin II is given. In transplantable mouse tumours the shutdown of the vasculature is also a limiting factor. Phthalocyanines of mixed sulphonation shut down the microvasculature during PDT as rapidly as porphyrins. A significant hypoxic cell fraction is created during light delivery and, no matter how much drug is present, or how much light is given, PDT will not be effective.

Jori: We obtain excellent tumour response if we use less than 1 mg/kg of liposomal Zn-Pc. At higher doses less necrosis is induced by PDT. Thus for effective use of phthalocyanines one must inject very low doses.

Henderson: In our tumours it doesn't work that way. Below 5 mg/kg we suddenly see no tumour response. There are other factors involved which we don't understand.

Kessel: How important is the time between administering the drug and irradiation? The optimum interval may not be the same for each dye.

Henderson: That is correct; if there are large amounts of drug in the circulation more vascular damage is produced. We do our tumour response experiments 24 h after injection when the dye has already accumulated in tissue.

Kessel: Is that the best time? Dr van Lier, did you irradiate at the same time after injection for all the phthalocyanines?

van Lier: We irradiated after 24 h because at that time in our mouse models the lipophilic dyes are at maximum concentrations in the tumour. However, pharmacokinetic studies with ^{14}C-labelled GaPcS indicate that the more water-soluble tri- or tetrasulphonated dyes reach a maximum tumour concentration 48 h after injection.

Kessel: Perhaps if you optimize the time you might see different results. We found that with several purified tetraphenylporphyrin sulphonates we could entirely miss the photodynamic effect if we picked the wrong time to irradiate. For some, the best time was 4 to 8 h after administration; after 24 h we wouldn't see any effect.

Dougherty: That is probably because one must use those drugs when the serum levels are high, because they are cleared quickly from tumours. There is then a risk of damage to everything in the light field. Therefore this may not be a viable option.

van Lier: With the highly sulphonated metallo-Pc dyes there is almost no material left in the bloodstream 48 h post-injection while tumour uptake reaches a maximum at this time. However, most of the dye localizes at non-vital sites within the tumour region.

Moan: In vitro studies show why the phthalocyanines don't work as efficiently as one would expect. The quantum yield of inactivation of cells for phthalocyanines is about 10% that for Photofrin II. Phthalocyanines probably localize in lysosomes, where damage may be less fatal for the cell than damage at other structures, such as mitochondria and plasma membranes, where Photofrin II localizes.

Dougherty: Do the fluorescence experiments show where in the cell the dyes are localized?

van Lier: In V-79 cells we can see intracellular structures. However, we have not attempted to recognize subcellular targets.

Moan: It depends on the amount of dye applied. With a sensitive detection device you can find it in lysosomes (Moan et al, this volume).

van Lier: One should not generalize about *in vitro* and *in vivo* distributions of phthalocyanines. Although these dyes are extremely potent singlet oxygen generators, the site and extent of biological damage depends on where they localize and this varies considerably with the type and the degree of substitution of the Pc macrocycle.

Bonnett: We place too much emphasis on the photophysical properties of sensitizers. We look for absorption coefficients, triplet lifetimes and singlet oxygen yields, but many compounds in this series have similar values for these parameters. The important differences between the photosensitizers, such as where they localize, are determined by their physical chemistries, particularly their solution and partition properties.

As we move to lower energies, at what stage do these photosensitizers cease to sensitize the formation of singlet oxygen? Is the reaction unfavourable for naphthalocyanines?

van Lier: The energies of the T_1 excited states of Al- and Zn-naphthalocyanines are very close to the energy required for singlet oxygen production. However, homogeneous solutions of these dyes and tryptophan gave, upon exposure to red light, good yields of typical $^1O_2$-tryptophan degradation products.

Dougherty: Is there an inherent relationship between the absorption wavelength and the energy of the triplet state which is important for singlet oxygen formation?

Truscott: Not necessarily. It depends on the energy gap between the S_1 state, produced by absorption, and the T_1 state. The energy of the T_1 state must be similar to or greater than that required to excite ground state oxygen to singlet oxygen (94 kJ/mol). It seems that for most porphyrins or phthalocyanines the S_1 and T_1 gap is of the order of 50–60 kJ/mol. Therefore you have to excite at about 800 nm maximum.

Dougherty: Does that mean that we should not look for dyes which absorb any further towards the infrared?

Bonnett: Has anyone looked at the next member out, anthracenocyanine?

van Lier: No, not to my knowledge.

Truscott: For compounds of that ring type, the S_1 to T_1 energy gap would probably be rather similar. If the absorption is at 900 nm the S_1 energy is about 130 kJ/mol which, if we subtract about 60 kJ/mol, leads to a T_1 energy below that required for singlet oxygen production. However, there may be other types of molecules for which the S_1 to T_1 gap is much smaller. For these we could look for longer wavelength absorption.

Morgan: We are only talking about singlet oxygen production. As we have discussed, singlet oxygen hasn't been detected *in vivo* and may not be the important agent in the mechanism.

Dougherty: I'd hate to bet against it at this stage of the game!

van Lier: Even if the photochemical mechanism during PDT involved a type I reaction, oxygen would be necessary to proliferate the biological damage.

Brown: We use tryptophan fluorescence quenching, that is degradation of tryptophan, as a possible test for singlet oxygen sensitization. My understanding is that this is not absolutely diagnostic, but some people think otherwise. Dr van Lier suggested that whereas identification of the hydroperoxide product is diagnostic, disappearance of tryptophan or identification of the N-formyl product is not diagnostic. Is this correct?

van Lier: Yes. Different products result from treating tryptophan with peroxy radicals or singlet oxygen. With peroxy radicals you get a dicyclic peroxide, whereas with singlet oxygen you get tricyclic pyrroloindole hydroperoxides

(Langlois et al 1986). These products can be distinguished by HPLC, but they all decompose to the same stable end products.

Bonnett: Cholesterol is the other marker; one looks for the 5-α-hydroperoxy derivative.

van Lier: Cholesterol is much less reactive towards singlet oxygen than tryptophan. Furthermore, 5-α-hydroperoxycholesterol rearranges to the 7-α-hydroperoxide which is characteristic for ground-state oxygen attack on cholesterol. It should be possible to use the 5-α-hydroperoxide derived from ^{14}C-labelled cholesterol, or hydroperoxy products derived from ^{14}C-labelled tryptophan, as a probe for $^1O_2$ if you can quench the reaction at an early stage and extract and analyse the hydroperoxides by chromatography.

Jori: We have studied the haematoporphyrin-sensitized photo-oxidation of tryptophan. One must take into account the sensitivity to the environment of the rate constant for the reaction of singlet oxygen with tryptophan. For example, this rate constant is five times lower in methanol than in aqueous solution. A pure singlet oxygen mechanism is followed in both media, but if one were to compare the rates at which tryptophan disappears one might believe that in methanol the singlet oxygen generation is much less efficient, which is not true; the quantum yield of singlet oxygen generation is the same, or perhaps higher, in methanol compared with in water.

Truscott: I suspect that the only unambiguous way to detect singlet oxygen is to look for its near-infrared luminescence.

Dougherty: Not even that would be definitive.

Truscott: Several people have said that phthalocyanines are good generators of singlet oxygen. This is true, but they are not as good as the porphyrins. Singlet oxygen yields for porphyrins are typically about 0.7 and most (but not all) phthalocyanines I have studied have singlet oxygen yields of 0.3 or less.

Jori: We found the quantum yield for singlet oxygen generation by zinc phthalocyanine to be 0.5 in methanol and 0.4 in liposomes.

Dougherty: The aggregation states of the dyes are important when we study these phenomena. We need to use the same solvent systems.

Moan: I think everybody agrees that the tryptophan furfuryl alcohol assay is not diagnostic for singlet oxygen in heterogeneous systems which contain lipophilic sensitizers. For water-soluble sensitizers, would a D_2O effect of about 10 (ten times faster degradation of tryptophan) be diagnostic for singlet oxygen?

Truscott: Most people would say that is diagnostic. There are examples, which I don't understand, of a D_2O effect on the triplet yield. That throws some doubt on such an assay, but normally there is a 10-fold increase in singlet oxygen-mediated tryptophan oxidation in D_2O.

Moan: What about the triplet lifetimes of these porphyrins and phthalocyanines in D_2O rather than H_2O?

Truscott: I have not measured that.

Jori: This is limited to aqueous media, not mixed solvents. It would be interesting to study the effect of D_2O on the type I mechanism. It might be an accelerator.

Lim: Dr van Lier, you said that in the *in vivo* studies the mice died when treated with high doses of phthalocyanines. Does that occur whether or not the animals are irradiated?

van Lier: These were not necessarily high doses—we used up to 2 mg/kg $ZnPcS_2$ and observed systemic toxicity with a red light fluence of 170–200 J/cm^2. No toxicity was observed in the absence of light. Up to 20 mg/kg $ZnPcS_4$ could be administered without inducing, in the presence of light, systemic toxicity. However, $ZnPcS_4$ also produced a much smaller PDT response in the experimental tumours in mice.

Lim: Do the mice die immediately after irradiation?

van Lier: No; this usually takes one to two days.

Lim: Have you tried to manipulate their inflammatory system by using cortisone or indomethacin?

van Lier: We used aspirin and antihistamine, but with little success. We did not try indomethacin.

Dougherty: C. J. Gomer has reported that for Photofrin II systemic toxicity was completely prevented by anti-inflammatory agents such as indomethacin (Abstract C-166, 2nd Congress of the European Society for Photobiology, 1987). This toxicity is a general shock phenomena resulting from damage to a large amount of tissue. It cannot be extrapolated to larger animals, such as rats.

Schieweck: The more lipophilic mono- and disulphonated metallo-Pc complexes are excreted via the liver. How long does complete excretion take? Have you studied liver toxicity?

van Lier: No, we have not. We followed the pharmacokinetics of ^{14}C-labelled sulphonated Ga-Pc for up to three weeks. During this period $GaPcS_2$ levels remained high, whereas only one third of the more hydrophilic $GaPcS_3$ remained after three weeks.

Dougherty: Does administering zinc phthalocyanine in liposomes damage the liver?

Jori: We didn't do toxicity studies. Some liposome-delivered zinc phthalocyanine remains in the liver of mice ten days after injection. It's mostly excreted through the faeces. The tetrasulphonated phthalocyanines are accumulated in large amounts by the kidneys, but they are not excreted in urine. We kept mice in metabolic cages for about two weeks and these dyes were still excreted through the faeces. I suspect that some renal toxicity might result from this prolonged retention and I do not understand why it happens.

Carruth: I have heard reports about retention of phthalocyanines in the eye.

van Lier: For the past year we have been collecting the eyes from mice treated by our Pc protocols for others who are looking for photodamage. The extent to which phthalocyanines are retained by the eye probably depends on the substituents on the macrocycle. However, we observed no significant differences in the retention of di- and tetrasulphonated Ga[^{14}C]PcS.

Dougherty: Have you done toxicity studies on the eye?

van Lier: No.

Bown: What about porphyrins and the eye?

Dougherty: The reports tend to be old and unclear. In one study haematoporphyrin derivative (HpD) was given to rabbits. Their eyes were kept open and a bright light was shone on them, but the effect of light alone was not studied. C. J. Gomer studied the distribution of HpD in the eye, and found it in the vascular regions, as expected. The retina can be damaged, but HpD is excluded from all other structures of the eye.

Morgan: Roberts & Dillon (1985) reported cross-linking of the crystalline lens proteins after *in vivo* irradiation.

Dougherty: Some patients treated for eye tumours by photodynamic therapy have formed cataracts.

Hönigsmann: As far as I know, no eye changes have been reported in erythropoietic protoporphyria (EPP) where large amounts of protoporphyrin circulate in the skin and the patients are highly photosensitive.

Bonnett: What about diseases with high levels of circulating uroporphyrin and coproporphyrin?

Hönigsmann: It's not clear whether erythropoietic coproporphyria really exists. But elevated uroporphyrin is found in Günthers disease. I don't think eye problems have been reported, but it is a rare disease.

Lim: Uroporphyrin is elevated in porphyria cutanea tarda (PCT), the most common porphyria. Although they have not been studied systematically, eye changes are not mentioned in the case reports.

Dougherty: Are porphyric patients sensitive to sunlight?

Lim: In EPP, redness and swelling follow exposure to sun. Patients with PCT develop skin fragility and blistering, but only in sun-exposed areas. Thus the sun does play a role, even though most patients do not complain of burning sensations on exposure to sunlight.

References

Langlois R, Ali H, Brasseur N, Wagner JR, van Lier JE 1986 Biological activities of phthalocyanines. IV. Type II sensitized photooxidation of L-tryptophan and cholesterol by sulfonated metal phthalocyanines. Photochem Photobiol 44:117–123

Moan J, Berg K, Kvam E et al 1989 Intracellular localization of photosensitizers. In: Photosensitizing compounds: their chemistry, biology and clinical use. Wiley, Chichester (Ciba Found Symp 146) p 95–111

Roberts JE, Dillon J 1987 *In vitro* studies on the photosensitized oxidation of lens proteins by porphyrins. Photochem Photobiol 46:683–688

General discussion

Comparison of haematoporphyrin derivative and new photosensitizers

van den Bergh: In the past five years we have treated about 50 cases of early cancers of the lungs, oesophagus and pharynx at the CHUV hospital in Lausanne. We find that the porphyrin mixtures (haematoporphyrin derivative, HpD) are fairly efficient photosensitizers, but the main problems are stenosis, stricture and fistula which apparently result from lack of selectivity. We are therefore searching for a new sensitizer that is more selective.

Dougherty: Our results, and those of Oscar Balchum, with Photofrin II (purified HpD) do not suggest that stenosis is a problem in photodynamic therapy (PDT) of lung cancer. Perhaps you should use less drug. We are investigating doses of 0.5 mg/kg Photofrin II for lung cancer. For skin cancers, where we have more data, the use of lower doses of Photofrin II greatly enhances the therapeutic effect, largely because of the photobleaching phenomenon.

van den Bergh: Have you looked at the rate of recurrence in the studies by Balchum et al (1984)?

Dougherty: That's being done now. We have our own data for patients treated several years ago. Photofrin II is perfectly safe, if used properly, in these patients with lung cancer. There is a greater problem with its use in the oesophagus.

Carruth: Bronchial strictures are not common but tracheo-oesophageal fistulas do occur.

Dougherty: It is true that the selectivity of a photosensitizer is important, but for a photosensitizer that doesn't photobleach an extremely high selectivity would be necessary to match the therapeutic ratio obtained with one that does.

Truscott: There's now evidence that freshly made tumour-localizing agents with ester-linked porphyrins become a mixture of esters and ethers. Has anyone compared the therapeutic effects of ester and ether-linked porphyrins?

Dougherty: We have used freshly prepared HpD, but I don't know what happens when it is injected into the animals. Therapeutically, it worked essentially the same way.

Kessel: Problems arise when you use an undefined mixture of unknown origin and unknown composition, which is all that is available for clinical use at present. We have separated up to 19 fractions from HpD. They all localize differently and have different effects in mice. It is not possible to compare results obtained with materials made in different institutions and stored differently.

Dougherty: We've studied the composition of Photofrin II by preparing all the isomers individually and analysing them by mass spectrometry. The major component of Photofrin II is the trimer with ether linkages, both with and without dehydration of the residual hydroxyethyl groups.

Truscott: Can you detect the larger polymers by mass spectrometry?

Dougherty: By looking hard enough we could detect up to octamers. But these were not significant.

Bonnett: It is increasingly difficult to volatilize oligomers as the molecular mass increases. It is therefore difficult to judge the composition of a mixture of oligomers by ordinary mass spectrometry.

What is the shelf-life of HpD or Photofrin II?

Dougherty: The shelf-life currently suggested for Photofrin II is one year if it is kept frozen and in the dark. It is not difficult to freeze the drug when you receive it, but it is a problem during shipment. The answer might be lyophilized drug, which will be supplied soon; its shelf-life is expected to be much longer and it probably doesn't require refrigeration, but I don't have any data on that.

Brown: Are you sure there is no change in the drug during the freeze–drying process?

Dougherty: That is being examined.

Wilson: How critical is the resolubilization procedure for the lyophilized drug? If you make small changes in the conditions do you get significantly different material?

Dougherty: I don't know.

Jori: Do we really think we shall find a sensitizer with better selectivity? Many people have been searching for a long time for a 'magic bullet' that targets tumours and not the surrounding tissue.

Bonnett: It is certainly worthwhile to look. We may not succeed, but I suspect we will. I think it is gradually emerging how we can better approach the search.

Berenbaum: There is no question that more selective sensitizers do exist—we already have them (Berenbaum et al 1986).

Kaye: As well as being more selective, the drug must be as active. Many new, more selective sensitizers are less active than the original porphyrins. I am still unsure whether if you use the optimum conditions for phthalocyanine you get better tumour kill than with Photofrin II.

Bown: How do you define 'better tumour kill'? Do you mean so many millimetres of necrosis within a tumour? Do you mean a form of tissue death that heals safely? Do you mean that when you put the same light dose on tumour and normal tissue only the tumour is damaged?

Kaye: Optimizing the therapeutic index so that you get maximum tumour kill without injury to normal surrounding tissue, can you kill more tumour with phthalocyanines than with Photofrin II?

Henderson: Only the clinical data will tell you that. In animals, with the mixture of phthalocyanines that most people use, I think the answer is no. But it may be possible with specific phthalocyanines or with different delivery systems.

Jori: To make a meaningful comparison you should perform a complete pharmacokinetic study for each photosensitizer because the time dependency of photosensitizer localization in the tumour varies according to the chemical structure, the carrier used and the type of tumour. As David Kessel said (p 28), the optimal irradiation time might be eight hours after injection for one sensitizer, but 24 hours for another.

Kaye: But I am saying to be as specific as you want; use optimum conditions, including time, for any specific phthalocyanine photosensitizer. Does that give better results than Photofrin II? I am not asking you to generalize about them.

Morgan: Steve Selman (1986) found significant responses in rat bladder tumours using tetrasulphonated chloroaluminium phthalocyanine at doses at which HpD produced no effect at all. A 50% reduction in tumour volume required 15 mg/kg of HpD.

Bown: It must depend on the tumour size. A figure such as 50% reduction in tumour volume is meaningless.

Berenbaum: We are discussing whether there is any way of measuring the selectivity of different sensitizers. This means comparing damage to tumour and to various normal tissues, where the sorts of damage observed may be qualitatively very different and therefore not measurable on the same scale. The answer is provided by using dose–response curves for the different types of damage to compare cost in damage to normal tissue with benefit in damage to the tumour. An example of this procedure is illustrated in Figs 1 to 3.

Figure 1 shows the dose–response curves for damage to tumour, skin and muscle for Photofrin II in the mouse. The measure of tumour damage here is depth of necrosis and the measure of skin and muscle damage is the acute oedema that develops after illumination (Berenbaum et al 1986, E. B. Chevretton, unpublished work). Using these superimposed curves, it is easy to answer the question, 'what cost in terms of skin or muscle damage is paid for a given benefit in terms of tumour necrosis?'. For example, for 3 mm depth of tumour necrosis, the cost in muscle damage is an increase of 45% in weight because of oedema, and the cost in skin oedema is a doubling in weight.

Figure 2 shows cost–benefit analyses obtained in this way for muscle oedema versus tumour necrosis for Photofrin II, tetra(*m*-hydroxyphenyl)porphyrin and tetra(*p*-hydroxyphenyl)porphyrin (*m*- and *p*-THPP), and tetra(hydroxyphenyl)-chlorins (*m*- and *p*-THPC). For 3 mm tumour necrosis, the cost in muscle damage is a 45% increase in weight for Photofrin II, 35% for *p*-THPP, 7% for *p*-THPC, 4% for *m*-THPP and 2% for *m*-THPC. Thus, selectivity for tumour versus muscle is least for Photofrin II and greatest for *m*-THPC. Figure 3 shows a similar analysis for skin oedema versus tumour damage, and again Photofrin II is the least selective sensitizer and *m*-THPC the most selective. For the latter, the disparity in sensitivity between tumour and skin was so great that

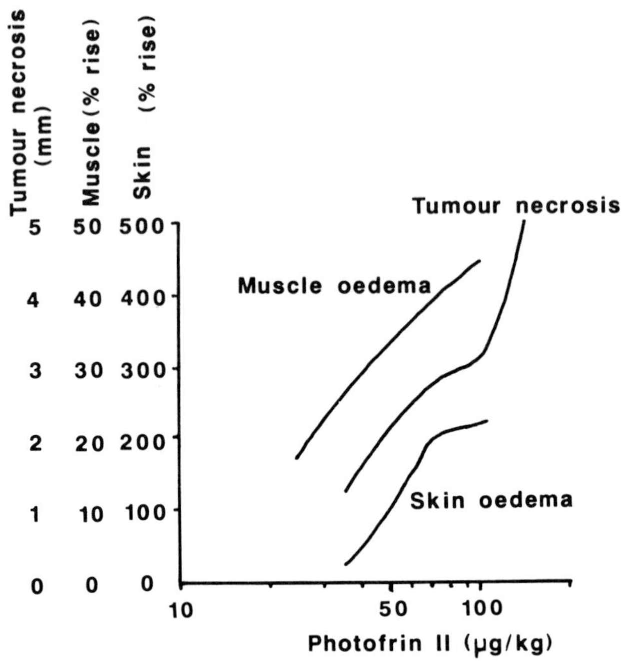

FIG. 1. (Berenbaum) Dose–response curves for tumour necrosis, skin oedema and muscle oedema in mice sensitized with Photofrin II. Tumour (subcutaneous PC6) and muscle (posterior aspect of leg) were illuminated one day and skin was illuminated three days after sensitization. The light dose was 10 J/cm^2 at 625 nm for all tissues. Depth of tumour necrosis was measured as described in Berenbaum et al (1982). Skin oedema was measured 4 h after illumination by weighing a 1 cm disc of illuminated skin (Berenbaum et al 1986), and muscle oedema (also at 4 h) by weighing the dissected posterior compartment leg muscles.

the dose–response curves overlapped at only one point, and nearly 6 mm depth of tumour necrosis (with a light dose of 10 J/cm^2) was obtained with almost negligible skin oedema.

With analyses of this sort, we avoid the difficulty that different tissues show different types of damage that cannot be directly compared with each other quantitatively, and it becomes a simple matter to compare the selectivity of various photosensitizers.

Brown: You cannot completely ignore other factors, such as time. For example, in the superficial treatments which we and others have done, a week after treatment there are nasty looking skin eschars and the therapeutic ratio seems very poor. But after two months the skin is healed and the tumour has gone.

Berenbaum: Any clinically relevant factor, including the timing of effects, can be incorporated into the cost–benefit method of measuring selectivity.

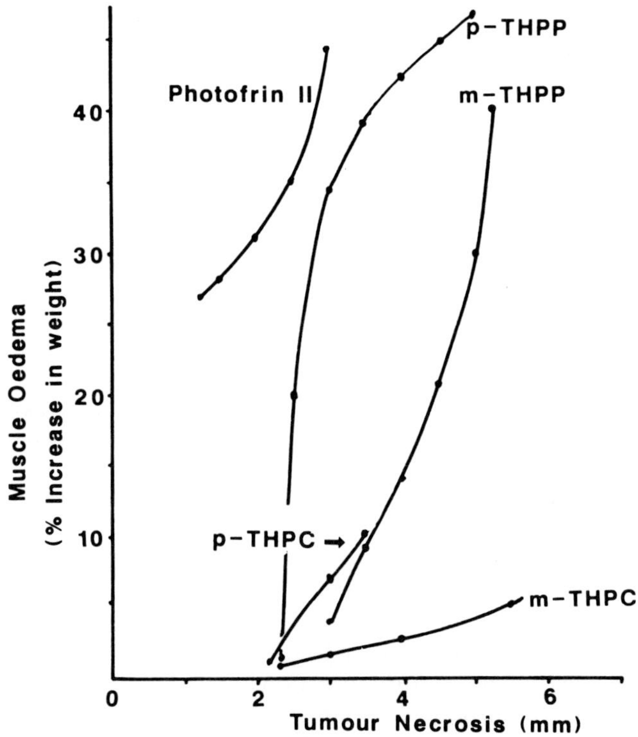

FIG. 2. (Berenbaum) Cost–benefit analyses for muscle oedema versus tumour necrosis for Photofrin II, tetra(*m*-hydroxyphenyl)porphyrin (*m*-THPP), tetra(*p*-hydroxyphenyl)-porphyrin (*p*-THPP) and the tetra(hydroxyphenyl)chlorins, *p*- and *m*-THPC. Cost in terms of muscle oedema is plotted against benefit in terms of tumour necrosis. These curves show the different selectivities for muscle versus tumour for the different sensitizers (E. B. Chevretton, unpublished work).

van den Bergh: This is not the measure of selectivity that one necessarily wants. When we treat a cancer in the oesophagus, we want to know what's happening right next to it, in the healthy part of the oesophagus, and not what's happening in the skin.

Berenbaum: A cost–benefit analysis can be done for any tissue where there is quantifiable damage. The results in Fig. 2 were for striated muscle, so they are relevant to the oesophagus.

Jocham: This provides one method of measuring selectivity. Nevertheless, you cannot extrapolate data from animal experiments to clinical situations. For example, there are distinct differences in the relative resistance to PDT of the urothelium of rats, of dogs and of man. One cannot draw conclusions before studying patients. The relative resistance of muscle will not be the same as that

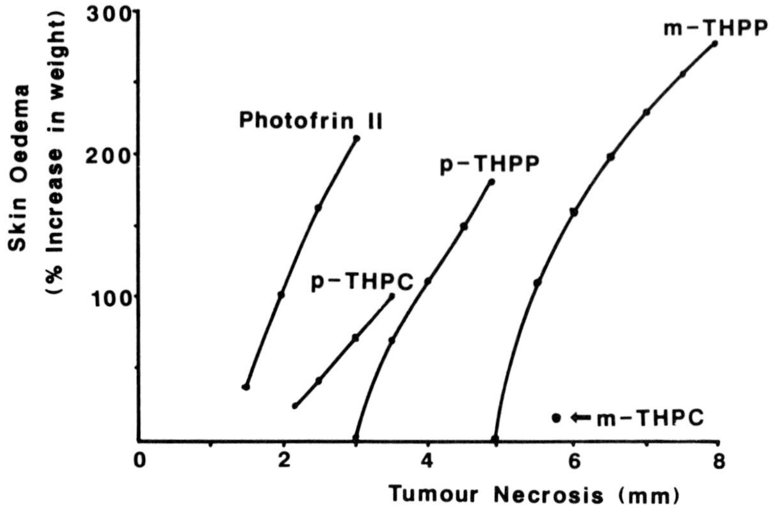

FIG. 3. (Berenbaum) Cost–benefit analyses for skin oedema versus tumour necrosis for Photofrin II, *m*- and *p*-THPP, and *m*- and *p*-THPC (S. L. Akande, unpublished work). See legend of Fig. 2 for abbreviations.

of the normal tissue in the tumour-bearing organ. In urinary bladder, for example, one must compare normal urothelium with the tumours.

Dougherty: But the principle is right. We need some selection process for sensitizers before clinical application. This provides a good clue.

Jori: What determines selectivity? Some closely related structures show very different selectivities according to this analysis.

Berenbaum: We think this is mainly a question of localization. *m*-THPP and *p*-THPP have almost identical singlet oxygen quantum yields (0.57 and 0.56 respectively) and their other photophysical properties are similar (Bonnett et al 1987, Berenbaum & Bonnett 1989). The illumination wavelengths used for the THPPs and THPCs are in the narrow range 648–656 nm. Their different selectivities cannot be accounted for by photophysical differences. Differences in localization are likely to be the governing factor, and this is a question of physical chemistry, not photochemistry.

Dougherty: The pattern of substituents in the chlorin *m*-THPC is identical to the one available in the porphyrin *m*-THPP. Why is the chlorin more selective than the corresponding porphyrin?

Bonnett: I think this is the first time such a comparison has been made (see Bonnett & Berenbaum, this volume). We do not fully understand the differences in selectivity, but we think they are related to the presence of hydroxyl groups.

Dougherty: But the porphyrins have hydroxyl groups as well.

Bonnett: Yes, but other factors have changed to modify the behaviour. Thus *m*-THPC does not have a centre of symmetry, whereas *m*-THPP does. The chlorin is expected to be less basic than the corresponding porphyrin, which will also have an important effect.

Dougherty: So simply hydrating a double bond has changed the selectivity.

Bonnett: It isn't hydrated; a double bond has been hydrogenated. The original reason for doing this was to increase absorption in the red.

Jori: But increasing absorption doesn't necessarily imply more selectivity.

Dougherty: No, but that's what happened. It seems to be a relatively minor chemical change and yet it made a major difference in terms of distribution. There must be a clue here.

Berenbaum: Calculations about how much light is absorbed cannot account for the different tissue selectivities. This is simply a question of the timing and location of the drug accumulation at the tissue, cellular and subcellular levels, which is determined by the physical chemistry and pharmacokinetics, as with any other drug.

Dougherty: But we should like to be able to predict that.

Bonnett: We shall achieve that by designing, synthesizing and testing more compounds. We are not clever enough yet to predict the magic bullet.

References

Balchum OJ, Doiron DR, Huth GC 1984 HPD photodynamic therapy for obstructing lung cancer. In: Porphyrin localization and treatment of tumours. Doiron DR, Gomer CJ (eds) Progr Clin Biol Res, Alan R Liss, New York, 170:727

Berenbaum MC, Bonnett R 1989 Tetra(hydroxyphenyl)porphyrins. In: Kessel D (ed) Photodynamic therapy of neoplastic disease. CRC Press Inc, Boca Raton, Florida, in press

Berenbaum MC, Bonnett R, Scourides PA 1982 *In vivo* biological activity of the components of haematoporphyrin derivative. Br J Cancer 45:571–581

Berenbaum MC, Akande SL, Bonnett R et al 1986 *meso*-Tetra(hydroxyphenyl)porphyrins, a new class of potent tumour photosensitizers with favourable selectivity. Br J Cancer 54:717–725

Bonnett R, Berenbaum MC 1989 Porphyrins as photosensitizers. In: Photosensitizing compounds: their chemistry, biology and clinical use. Wiley, Chichester (Ciba Found Symp 146) p 40–59

Bonnett R, Ioannou S, White RD, Winfield U-J, Berenbaum MC 1987 *meso*-Tetra(hydroxyphenyl)porphyrins as tumour photosensitizers: chemical and photochemical aspects. Photobiochem Photobiophys Suppl:45–56

Selman SH, Kreimer-Birnbaum M, Chaudhuri K et al 1986 Photodynamic treatment of transplantable bladder tumors in rodents after pretreatment with chloroaluminium tetrasulfophthalocyanine. J Urol 136:141–145

Porphyrins as photosensitizers

Raymond Bonnett and *Morris Berenbaum

Department of Chemistry, Queen Mary College, Mile End Road, London E1 4NS, UK and
**Department of Experimental Pathology, St Mary's Hospital Medical School, London W2 1PG, UK*

Abstract. The porphyrins have two important roles in photobiology: in photosynthesis, which has evolved and is highly organized morphologically; and in the photodynamic effect, which is adventitious. The damage to tissue that results from photodynamic action is regarded as arising from a number of pathways, but singlet oxygen generation is a major route. Even for the latter mechanism, a number of target molecules are possible. Compared with photosynthesis, it is a very disorganized process. Since the mid 1970s there have been increasing efforts to turn the photodynamic effect to good use as the basis of a phototherapy for cancer. The field has been dominated by the photosensitizer haematoporphyrin derivative. This is a complex mixture of molecules, and although we are learning more about it, it seems unlikely that it will be possible to separate out a useful single substance. A second generation of porphyrin sensitizers is emerging, which, unlike haematoporphyrin derivative, is designed for the job in hand. Some of these photosensitizers, especially the hydroporphyrins, look quite promising.

1989 Photosensitizing Compounds: their Chemistry, Biology and Clinical Use. Wiley, Chichester (Ciba Foundation Symposium 146) p 40–59

The porphyrin structure (1) can be thought of as a molecular tablet, approximately flat and about 7 Å square, with a hole in the middle which can accommodate a metal ion. It has a rich chemistry, and, associated with the striking absorption spectrum, an interesting photochemistry. The dihydroporphyrin system, or chlorin (2), has quite distinct chemical and physical properties. In particular, its absorption spectrum shows a strong band in the red region of visible light, giving the substance a green colour.

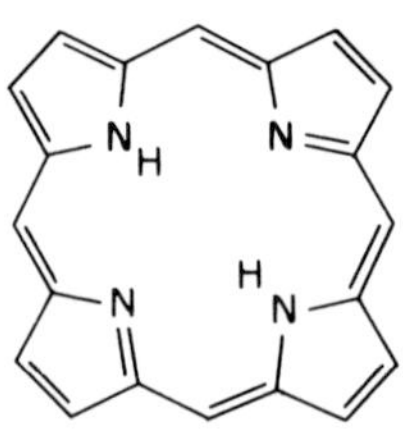

(1)

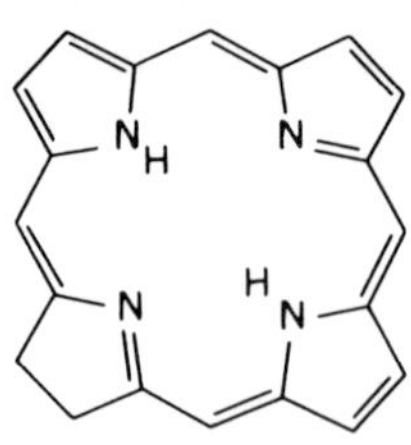

(2)

The fascinating chemistry of these compounds and their relatives changes subtly as the central metal is changed. The importance of this chemistry is highlighted by the natural occurrence and function of these compounds as the 'pigments of life', notably iron complexes of porphyrins (haemoglobin, myoglobin, cytochromes) and magnesium complexes of chlorins (chlorophylls). The most significant photosensitization for living things on this planet is photosynthesis. This provides essentially the total storable energy input into the biosphere, and depends mainly on two magnesium (II) dihydroporphyrins, chlorophylls *a* and *b*.

Photosynthesis is an *evolved* photobiological process which operates on a very short time-scale with few side-reactions because the components of the photoreaction centre have a precise and purposeful relationship to one another (Deisenhofer & Michel 1987). As illustrated in Fig. 1, the processes involved in tumour phototherapy are quite different. Tumour phototherapy is an example of a photodynamic effect in which the photosensitizer is an intruder in a tissue environment which is not set up to deal with the chemical reactions resulting from the generation of the electronically excited photosensitizer. Consequently the chemical mechanism of tumour phototherapy involves many reactions. The formation of singlet oxygen is commonly considered to be the major pathway, but several biological targets are available for this reactive molecule. Some of these, for example unsaturated lipid, cholesterol, and certain amino acid side-chains in proteins, are important membrane components, and photochemical changes in these are thought to be an important cause of tissue damage.

Such damage has for a long time found example in patients suffering from some of the porphyrias. However, the first proof that porphyrins were photodynamic agents in man was provided by Meyer-Betz (1913) who in a celebrated, if foolhardy, experiment injected himself with 200 mg of haematoporphyrin and became very sensitive to visible light. At about this time Fischer made a number of observations on the effect of porphyrin structure on photodynamic activity which are of some interest in the context of current work. Fischer & Meyer-Betz (1912) observed that whereas haematoporphyrin caused a pronounced photosensitization (subcutaneous administration in mice), mesoporphyrin did not (although the photobiological activities were apparently reversed in paramecia; Fischer & von Kemnitz 1916). On the other hand, uroporphyrin was found to be almost as phototoxic as haematoporphyrin (Fischer 1916). In retrospect these early studies, which Fischer does not seem to have continued after the First World War, illustrate the problems, which are still with us, of getting the porphyrin sensitizer into solution.

Haematoporphyrin derivative

In recent years the subject of porphyrin photosensitization has developed in two ways. In chemical studies, porphyrins are routinely used to generate singlet

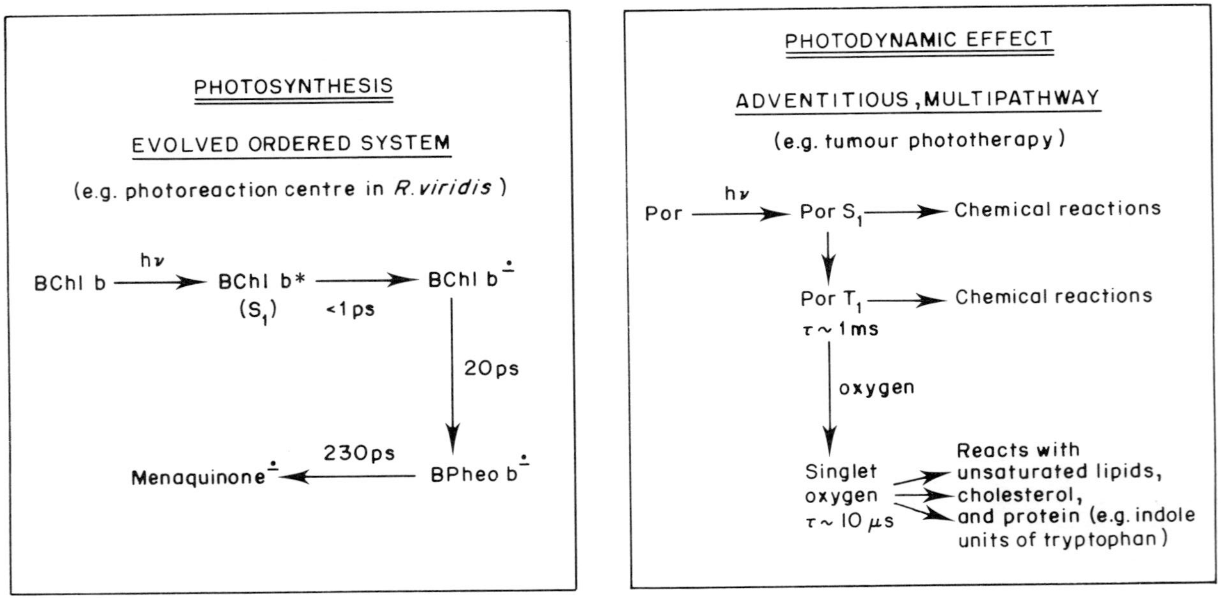

FIG. 1. Two distinct types of photobiological reaction involving light absorption by porphyrins. Photosynthesis occurs in a highly ordered system which has evolved to give a single biologically desirable result. Process lifetimes are short. Photodynamic action involves species with much longer lifetimes. These species are generated in a much less ordered bioenvironment, and many reaction pathways are possible. BChl b, bacteriochlorophyll *b*; BPheo b, bacteriopheophytin *b*; Por, porphyrin; S_1, first excited singlet state; T_1, first excited triplet state; $^3O_2$, ground-state triplet oxygen.

oxygen, particularly for synthetic work. Zinc(II) *meso*-tetraphenylporphyrin is often used for this purpose. When a water-soluble photosensitizer is needed, the tetrasodium salt of *meso*-tetraphenylporphyrin tetrasulphonic acid ($TPPS_4Na_4$) can be used (Bonnett & Ioannou 1986; Equation 1).

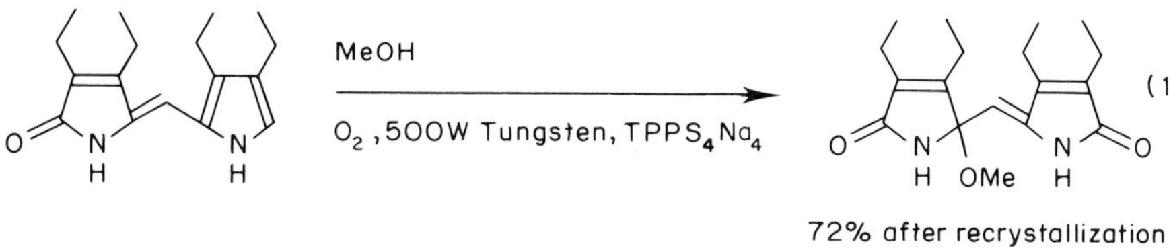

(1)

Biomedical studies with sensitizers have been dominated by haematoporphyrin derivative (HpD) which was discovered by Lipson & Baldes (1960). The steps involved in its preparation are summarized in Equation 2.

Haematoporphyrin (3) —[5% H_2SO_4 in HOAc; 30 min, RT]→ HpD (Stage I) —[treatment with base]→ HpD (Stage II) (2)

HpD (Stage II) ↓ Remove monomeric porphyrins → More active fraction e.g. Fraction D, commercial preparations

An important property of HpD is that it contains components which localize in tumour tissue in a way that allows the tumour to be visualized by fluorescence (Lipson et al 1961) or to be photodegraded (Diamond et al 1972).

HpD (Stage I) consists of a mixture of molecules in which vinyl, 2-acetoxy ethyl and 2-hydroxyethyl substituents are permutated at the 3 and 8 positions of deuteroporphyrin (Fig. 2): the major component is the diacetoxy derivative (4) (Bonnett et al 1981). Alkaline treatment is used to prepare the substance for administration (HpD Stage II), and this causes hydrolysis or elimination of the acetoxy groups to give monomeric porphyrins—haematoporphyrin (3), hydroxyethylvinyl deuteroporphyrin (5) and protoporphyrin (6). These monomers are not active in the *in vivo* biological assay with tumour implants in mice, but a higher molecular mass fraction, which we called Fraction D, is formed concomitantly and does retain activity (Berenbaum et al 1982). We proposed that this fraction contained dimers or oligomers with ether, ester, or carbon–carbon linkages (Fig. 2), all of which can be readily rationalized in terms of chemical mechanism.

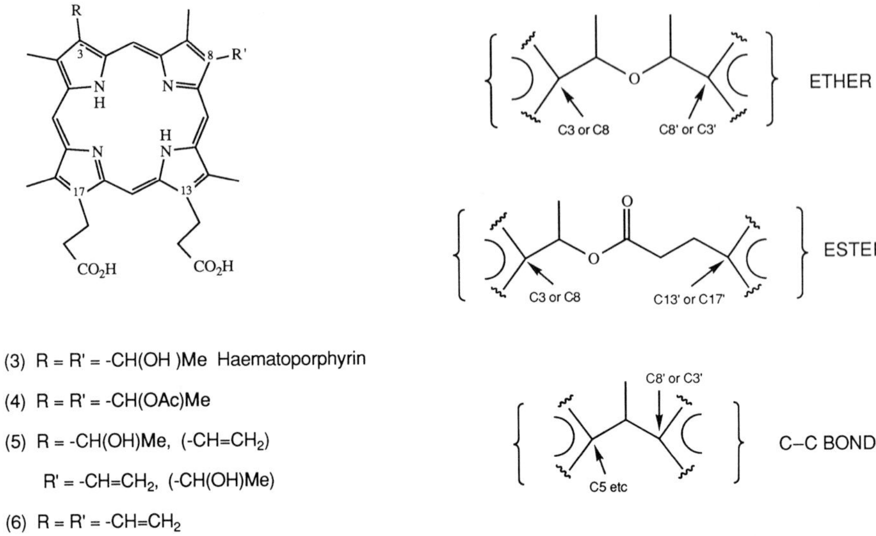

FIG. 2. Structures of haematoporphyrin and its monomeric and oligomeric congeners. The right hand side of the figure shows postulated interporphyrin linkages (Berenbaum et al 1982).

These proposals have led to a large amount of activity. Some workers have favoured the ether structure (Dougherty et al 1984, Scourides et al 1987), which has even appeared in a patent (Weishaupt et al 1984); others have favoured the ester structure, or a mixed ether–ester (Kessel 1986). There is some evidence for the initial formation of an ester under certain conditions with subsequent conversion to an ether (Byrne et al 1987). But it has not proved possible to isolate a highly active pure oligomer from the high molecular mass fraction of HpD; the complex pattern of products revealed by high-performance liquid (HPLC) or gel-permeation (GPC) chromatography and the various molecular mass values reported lead us to think that this is not a useful way forward.

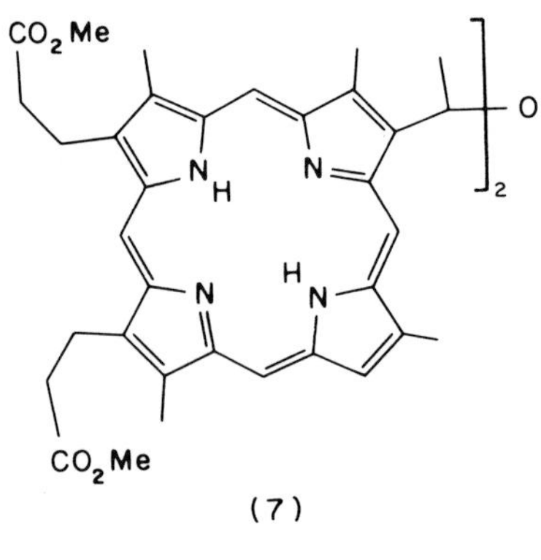

(7)

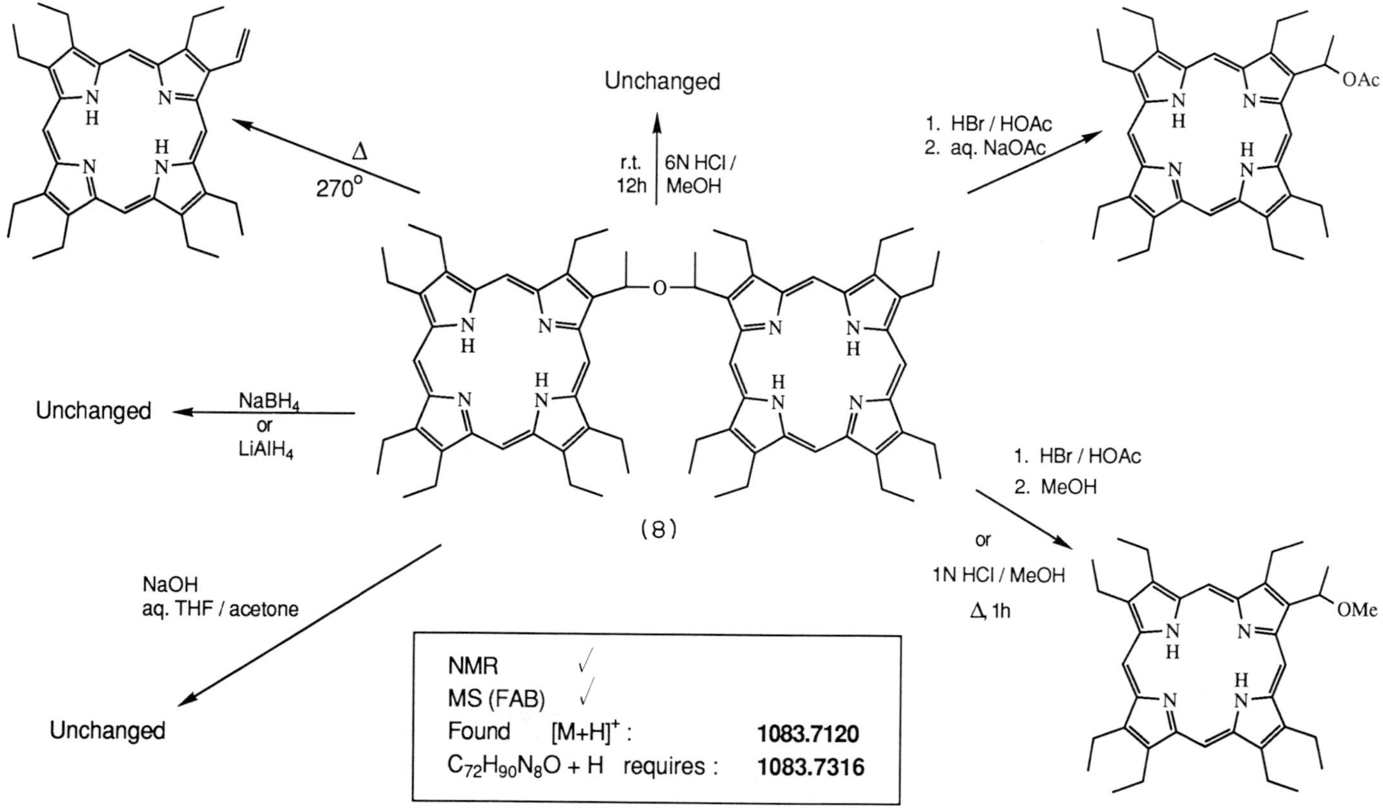

FIG. 3. Chemical reactions of a di(porphyrinylmethyl) ether (8).

Another approach has been to synthesize dimeric structures, although the positional and stereo isomers continue to cause problems (Scourides et al 1987, Morris & Ward 1988, Pandey et al 1988). Pandey & Dougherty (1988) have reported the production, by what is probably the best synthesis so far, of the ether (7) in 58% yield. It is, one must presume, a mixture of diastereoisomers, but is active in the *in vivo* biological assay.

With Drs D. Moffat and A. Nizhnik we have been interested in the chemical properties of such ethers, which might be unusual because of the large, highly aromatic nuclei flanking the ether linkages. We prepared the model ether (8) and examined its chemistry, which is summarized in Fig. 3.

Second-generation porphyrin photosensitizers

HpD, however presented, is a complex variable mixture. Therefore it is difficult to establish reliable dose–response relationships. For the past few years we have been selecting, synthesizing, and testing with the *in vivo* biological assay a large number of porphyrins. From our results, and from those in the literature, it is possible to deduce the desirable features for a tumour photosensitizer:

(a) It should be non-toxic in the dark;

(b) The pharmacokinetic behaviour should be such that it is selective for tumour tissue over surrounding normal tissues and is not retained in the body for long periods. This behaviour appears to be related to the water solubility and partition coefficient (lipid/water) of the substance, which render it amphiphilic;

(c) It should have constant composition, and preferably be a pure single substance;

(d) It should have a high triplet-state quantum yield, and a triplet-state energy greater than 94 kJ/mol with efficient energy transfer to produce singlet oxygen;

(e) It should absorb in the red part of the visible spectrum.

The last criterion arises because absorption and scattering of red visible light by human tissue is much less extensive than of blue light (Wan et al 1981). In other words, red light penetrates better than blue light; hence it is advantageous to be able to use red light for tissue photosensitization. There is currently much work on this aspect of the problem. Absorption in the red has been increased by: (a) modifying the substitution pattern (e.g. Bonnett et al 1987); (b) using 2,3-dihydro or 7,8,17,18-tetrahydroporphyrins (chlorin and bacteriochlorin systems respectively) (Selman et al 1987, Kessel & Smith 1989); or (c) more deep-seated changes in the structure to give, for example, phthalocyanine and related systems (e.g. Ali et al 1988). Figure 4 illustrates the effect of these changes on absorption in relation to tissue transmission.

We have used strategies a and b. Bioassay of a large number of porphyrins revealed that the tetra(hydroxyphenyl)porphyrin isomers (9,10,11) were effective tumour photosensitizers (Fig. 5). Moreover, the *meta* (10) and *para* (11) isomers

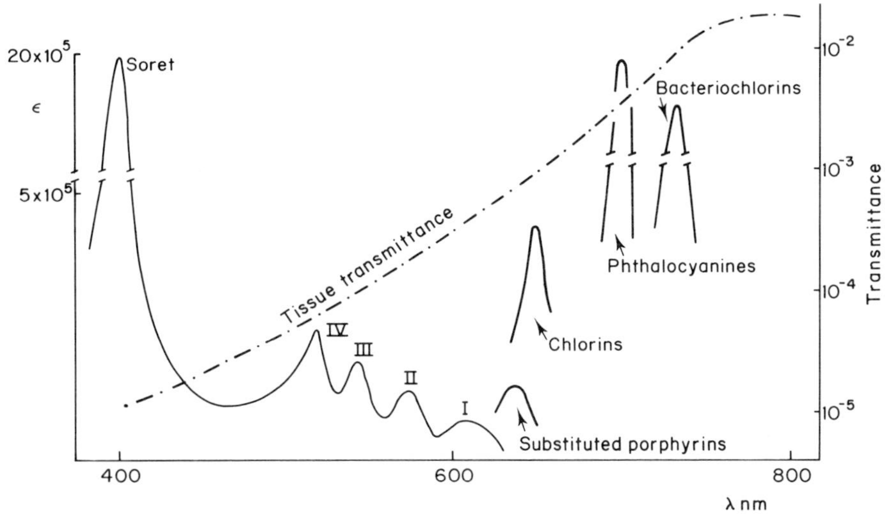

FIG. 4. Diagrammatic representation of the absorption spectrum of a typical metal-free porphyrin, and of the modifications caused by various structural changes, in relation to the transmittance of 1 cm of human tissue.

showed a good selectivity for tumour rather than muscle and skin (Berenbaum et al 1986). These compounds are both phenols and porphyrins: the hydroxyl function confers some degree of solubility in polar media, and at the same time has an auxochromic effect, shifting Band I from about 625 nm (e.g. in HpD) to about 650 nm, with an increase in molar extinction.

Reduction of these porphyrins with diimide gives the corresponding chlorins (12,13,14) (Fig. 5). In the *meta* series, the reduction has been carried as far as the tetrahydro stage, and the bacteriochlorin (15) has been characterized. The chlorins all show the typical strong band ($\epsilon \sim 20\,000$) at about 650 nm, whereas in the bacteriochlorin this band has, as expected, shifted further to the red and intensified so that it now appears at about 735 nm ($\epsilon \sim 90\,000$).

Bioassay shows that these hydroporphyrins are markedly more potent than the corresponding porphyrins (Table 1); the *para* and *meta* isomers cause substantial tumour necrosis at dose levels at which the corresponding porphyrins are ineffective. The bacteriochlorin (15) proved to be exceptionally active. This is the first time that a comparison has been made between sensitizers with a given substitution pattern at various levels of reduction derived from parent porphyrins of demonstrated efficacy.

Another interesting chlorin and bacteriochlorin system is based on octaethylporphyrin, which can be conveniently prepared in a pure condition and in appreciable quantity. As before, the aims of the work were to introduce substituents that confer amphiphilic properties and, at the same time, to shift the absorption into the red. Hydroxy functions have again been used to produce

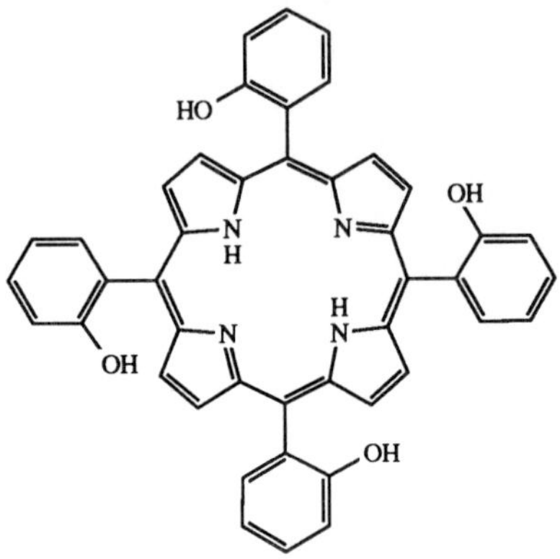

(9) 5,10,15,20-Tetra(*o*-hydroxyphenyl)porphyrin
o-THPP

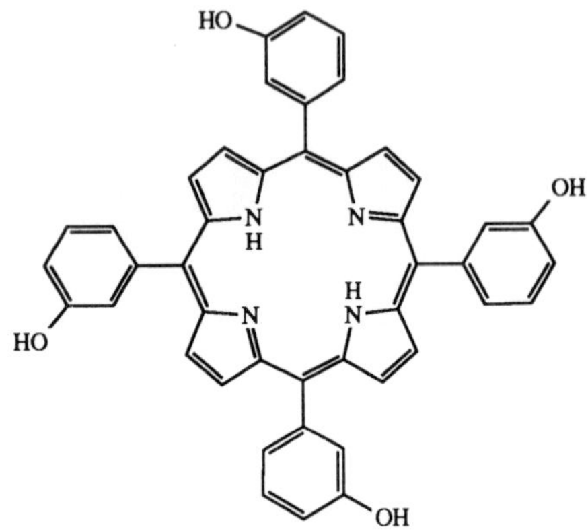

(10) 5,10,15,20-Tetra(*m*-hydroxyphenyl)porphyrin
m-THPP

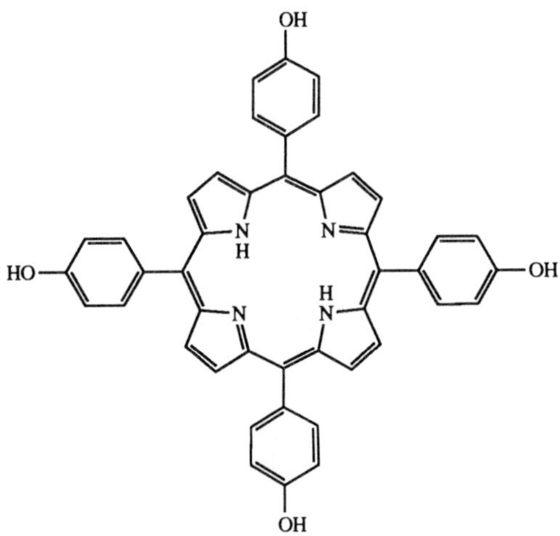

(11) 5,10,15,20-Tetra(*p*-hydroxyphenyl)porphyrin
p-THPP

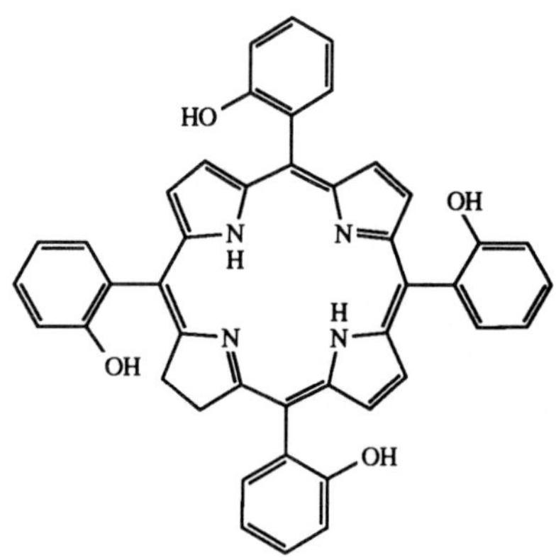

(12) 5,10,15,20-Tetra(*o*-hydroxyphenyl)chlorin
o-THPC

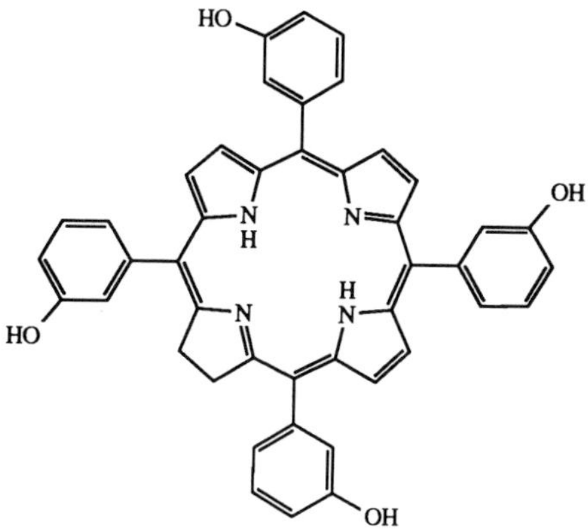

(13) 5,10,15,20-Tetra(*m*-hydroxyphenyl)chlorin
m-THPC

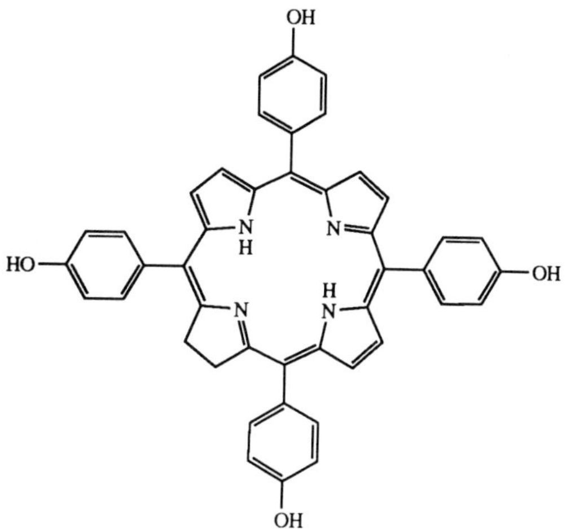

(14) 5,10,15,20-Tetra(*p*-hydroxyphenyl)chlorin
p-THPC

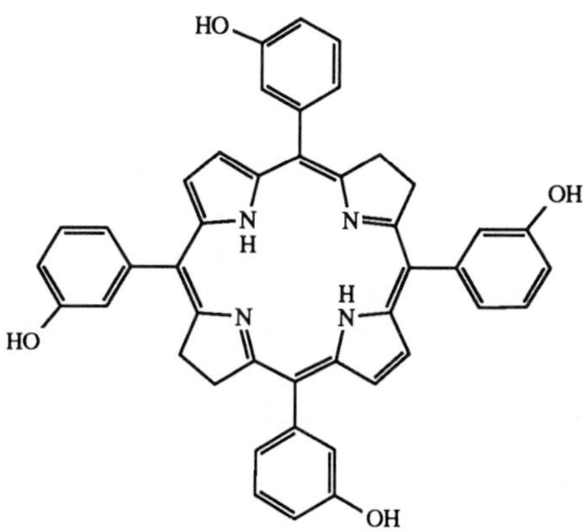

(15) 5,10,15,20-Tetra(*m*-hydroxyphenyl)bacteriochlorin
m-THPBC

FIG. 5. Porphyrins and reduced porphyrins in the *meso*-tetra(hydroxyphenyl) series.

TABLE 1 Tumour photonecrosis with hydroporphyrins

Photosensitizer[a]	*Dose (μM/kg)*	*Wavelength*[b] *(nm)*	*Depth of tumour necrosis (mm ± SE)*[c]
o-THPC (12)	6.25	652	4.16 ± 0.27 (14)
	3.125	652	3.69 ± 0.74 (9)
	1.562	652	0.31 ± 0.31 (8)
m-THPC (14)	0.75	652	5.41 ± 0.39 (19)
	0.375	652	3.79 ± 0.28 (6)
	0.2	652	0.13 ± 0.05 (12)
p-THPC (14)	6.25	653	3.50 ± 0.54 (10)
	3.125	653	2.13 ± 0.50 (10)
m-THPBC (15)	6.25	741	died within 2 h,
	3.125	741	tumours visibly
	1.562	741	blackening
	0.39	741	5.22 ± 1.21 (8)
Dihydroxy OEC (16)	6.25	645	6.29 ± 0.98 (7)
	3.125	645	2.88 ± 0.93 (10)
	1.562	645	0.86 ± 0.33 (9)
Trihydroxy OEBC (21)	25	713	4.00 ± 0.29 (6)
	6.25	713	1.75 ± 0.52 (6)

[a]THP, tetra(hydroxyphenyl); C, chlorin; BC, bacteriochlorin; OE, octaethyl. For structures see formula indicated by number.
[b]The total energy administered was 10 J/cm^2 throughout (Berenbaum et al 1982, 1986).
[c]Number of tumours in parenthesis.

these changes but the mechanism—β-addition—is quite different (Fig. 6). Treatment of octaethylporphyrin with osmium tetroxide gives the *cis*-diol (16) (Klotmann 1964). When treated with acid, this undergoes a pinacol–pinacolone change to the β-oxochlorin (17) (Bonnett et al 1964). Borohydride reduction gives the secondary alcohol (18) which, via the bromo compound, can be attached to polar fragments. For example, reaction with cysteine gives compound (19). All these molecules (16–19) are at the chlorin level, with absorption bands at about 650 nm. To enter the bacteriochlorin series, the β-oxochlorin (17) is subjected to a second stage of hydroxylation to give the β-oxobacteriochlorindiol (20) (Inhoffen & Nolte 1969, Chang et al 1986) which can be reduced to the trihydroxybacteriochlorin (21). That these compounds might be useful as tumour photosensitizers is shown by the results for the dihydroxychlorin (16) and the trihydroxychlorin (21) in Table 1. Moreover, we think that the possibility of increasing the number of polar groups (hydroxy, for example, in the series 18, 16, 21) in a stepwise fashion may facilitate adaptation of the treatment to differing tumour requirements.

In a sense, HpD was discovered by accident, but through it attention has been focused on the possibility of tumour phototherapy. HpD has several

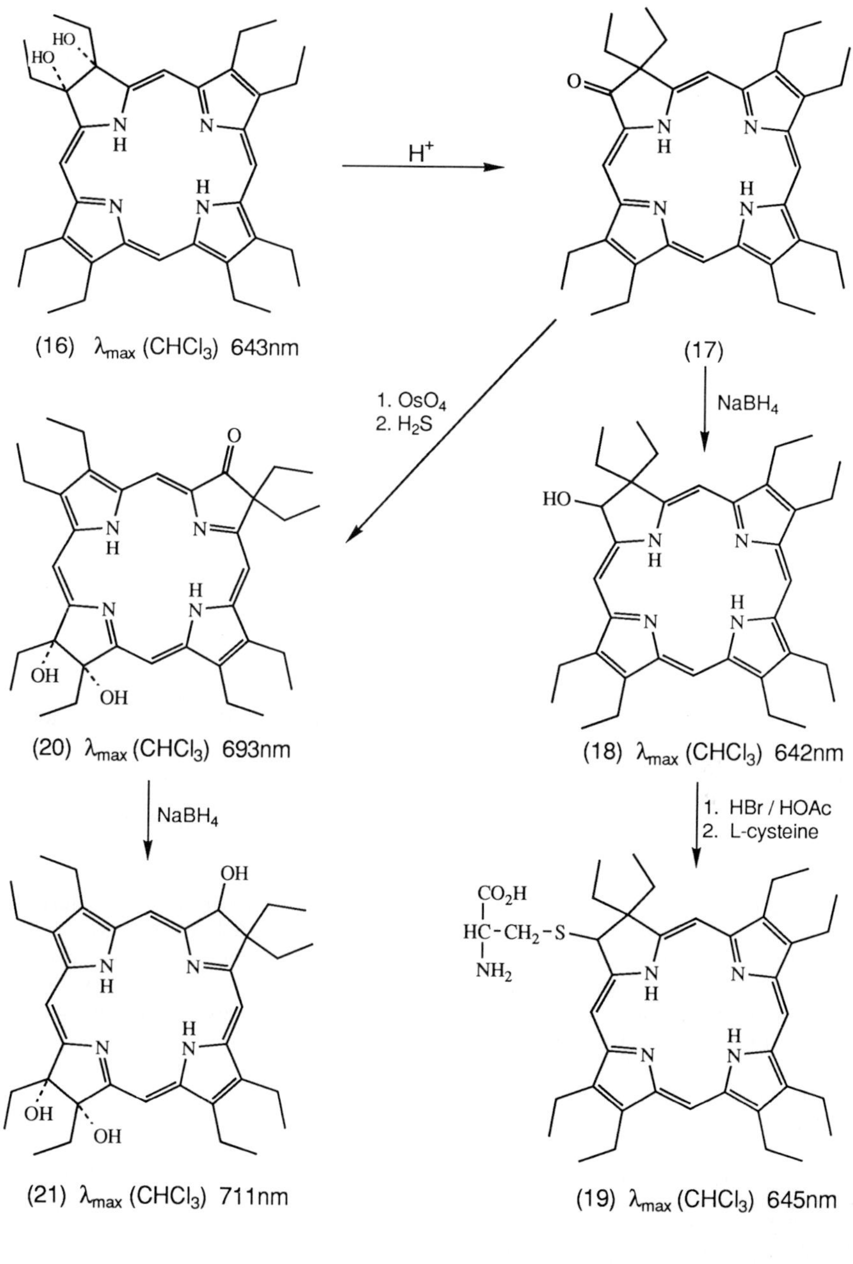

FIG. 6. Chlorins and bacteriochlorins with β-substitution.

disadvantages, but it would be ironic if the approach were to suffer because of the imperfections of the first effective drug. The increasing efforts being made worldwide on photosensitizer synthesis lead us to the optimistic view that a second-generation photosensitizer, not found by accident but purpose-built, will soon offer a viable alternative to HpD.

Acknowledgements

The new work referred to here was carried out by Miss S. L. Akande, Dr E. Chevretton, Dr G. Hall, Miss E. Lindsay, Dr D. Moffat, Dr A. N. Nizhnik, Dr R. D. White and Miss U. J. Winfield. We thank Scotia Pharmaceuticals Ltd, the Cancer Research Campaign, the London Hospital Special Trustees, and the British Council for support.

References

Ali H, Langlois R, Wagner JR, Brasseur N, Paquette B, van Lier JE 1988 Biological activities of phthalocyanines. X. Syntheses and analysis of sulphonated phthalocyanines. Photochem Photobiol 47:713–717

Berenbaum MC, Bonnett R, Scourides PA 1982 *In vivo* biological activity of the components of haematoporphyrin derivative. Br J Cancer 45:571–581

Berenbaum MC, Akande SL, Bonnett R et al 1986 *meso*-Tetra(hydroxyphenyl)porphyrin, a new class of potent tumour photosensitisers with favourable selectivity. Br J Cancer 54:717–725

Bonnett R, Ioannou S 1986 On the structure of the alkanol propentdyopents. J Chem Soc Chem Commun, p 213–214

Bonnett R, Dolphin D, Johnson AW, Oldfield D, Stephenson GF 1964 The oxidation of porphyrins with hydrogen peroxide in sulphuric acid. Proc Chem Soc p 371–372

Bonnett R, Ridge RJ, Scourides PA, Berenbaum MC 1981 On the nature of haematoporphyrin derivative. J Chem Soc Perkin Trans 1 p 3135–3140

Bonnett R, Ioannou S, White RD, Winfield UJ, Berenbaum MC 1987 *meso*-Tetra(hydroxyphenyl)porphyrins as tumour photosensitisers: chemical and photochemical aspects. Photobiochem Photobiophys (Suppl) p 45–46

Byrne CJ, Marshallsay LV, Ward AD 1987 The structure of the active material in haematoporphyrin derivative. Photochem Photobiol 46:575–580

Chang CK, Sotiriou C, Wu W 1986 Differentiation of bacteriochlorin and isobacteriochlorin formation by metallation. J Chem Soc Chem Commun p 1213–1215

Deisenhofer J, Michel H 1987 The structure of the photosynthetic reaction centre from *Rhodopseudomonas viridis*. Photobiochem Photobiophys (Suppl) p 133–140

Diamond I, McDonagh AF, Wilson CB, Granelli SL, Nielsen S, Jaenicke R 1972 Photodynamic therapy of malignant tumours. Lancet ii:1175–1177

Dougherty TJ, Potter WR, Weishaupt KR 1984 The structure of the active component of haematoporphyrin derivative. Adv Exp Med Biol 170:301–314

Fischer H 1916 On the toxicity, the sensitising action, and the spectroscopic properties of natural porphyrins. Hoppe-Seyler's Z Physiol Chem 97:109–127

Fischer H, Meyer-Betz F 1912 On porphyrin formation. Hoppe-Seyler's Z Physiol Chem 82:96–108

Fischer H, von Kemnitz GA 1916 On the action of some porphyrins on paramecia. Hoppe-Seyler's Z Physiol Chem 96:309–313

Inhoffen HH, Nolte W 1969 Oxidative transformation of octaethylporphyrin to geminiporphin-polyketones. Liebig's Ann Chem 725:167–176

Kessel D 1986 Proposed structure of the tumour-localising fraction of HpD. Photochem Photobiol 42:193–196
Kessel D, Smith K 1989 Photosensitisation with derivatives of chlorophyll. Photochem Photobiol 49:157–160
Klotmann G 1964 Hydroxylation of porphyrins and rearrangement of dihydroxychlorins. Dissertation, Braunschweig, p 55–73
Lipson RL, Baldes EJ 1960 The photodynamic properties of a particular haematoporphyrin derivative. Arch Dermatol 82:508–516
Lipson RL, Baldes EJ, Olsen AM 1961 The use of a derivative of haematoporphyrin in tumour detection. J Natl Cancer Inst 26:1–8
Meyer-Betz F 1913 Investigations on the biological (photodynamic) action of haematoporphyrin and other derivatives of blood and bile pigments. Deutsch Arch Klin Med 112:476–503
Morris IK, Ward AD 1988 The synthesis of dihaematoporphyrin ether and related porphyrin dimers. Tetrahedron Lett 29:2501–2504
Pandey RK, Dougherty TJ 1988 Synthesis and photosensitising activity of a diporphyrin ether. Photochem Photobiol 47:767–777
Pandey RK, Dougherty TJ, Smith KM 1988 Synthesis of haematoporphyrin dimers and trimers with ether linkages. Tetrahedron Lett 29:4657–4660
Scourides PA, Bohmer RM, Kaye AH, Morstyn G 1987 Nature of the tumour-localising components of haematoporphyrin derivatives. Cancer Res 47:3439–3445
Selman SH, Garbo GM, Keck RW, Kreimer-Birnbaum M, Morgan AR 1987 A dose response analysis of purpurin derivatives used as photosensitisers for the photodynamic treatment of transplantable urothelial tumours. J Urol 137:1255–1257
Wan S, Parrish JA, Anderson RR, Madden M 1981 Transmittance of non-ionising radiation in human tissues. Photochem Photobiol 34:679–681
Weishaupt R, Dougherty TJ, Potter WR 1984 Purified haematoporphyrin derivative for diagnosis and treatment of tumours. Intl Patent Publ No. WO 84/01382 to Oncology Research and Development Inc, US

DISCUSSION

Dougherty: Photofrin II is a completely reproducible material; it gives the same kind of biological efficacy each time it is used. The fact that it is a mixture is not an issue with the drug regulatory agencies.

The aim of making sensitizers that are activated by longer wavelengths is right, and in this context I wish to emphasize something raised by Dr van Lier (this volume). A practical advantage, beside the theoretical advantages, of working with drugs in the 800 nm region is that high power, simple light sources—diode lasers made by Spectra Diode—are already available. We have a 100 mW laser in our laboratory which Bill Potter has adapted for treatment of mice. It emits at 780 nm and we demonstrated, for example, its efficacy for activating bacteriochlorophyll.

We recently purchased a 1000 mW diode laser which is also a simple system. The availability of simple apparatus which doesn't require much power would make photodynamic therapy more acceptable to clinicians. The currently available equipment can only be used in a large cancer centre, because it is bulky and requires a technician. If we had a drug that was as effective as Photofrin II, all other things being equal, and could be used with a diode laser, it would immediately replace all other photosensitizers.

Truscott: Professor Bonnett, you made chlorins from porphyrins by reduction with diimide. How did you make the bacteriochlorin?

Bonnett: It's the same reaction. After a short time you get mainly the chlorin and if you leave it longer you get the bacteriochlorin. You still have to do some column chromatography after the reduction.

Berenbaum: The fact that a photosensitizing drug is taken up more in the liver than elsewhere does not mean that you can't use it to treat tumours in liver. There is much spare tissue in the liver. If you could destroy the tumour, even at the expense of the surrounding 2 or 3 cm of normal liver, that would be extremely useful.

Bonnett: I am glad to know that. Many drugs are prepared for excretion by the action of cytochrome *P*-450 in the liver. Thus porphyrins are likely to be present in normal tissue as well as in the tumour. How can one irradiate selectively?

Bown: One simply puts an optic fibre in the middle of the tumour. By PDT with haematoporphyrin derivative (HpD) and sulphonated aluminium phthalocyanine we produced necrosis in liver which healed completely, with regeneration of the liver to replace the destroyed part (Bown et al 1986).

Jori: Was this done on an infiltrating tumour?

Bown: We used normal livers, but if you are dealing with completely dead tissue it makes little difference whether it's liver or tumour; it heals the same way.

Henderson: Do you inflict the liver damage by insertion of a fibre or with external light?

Bown: We inserted a fibre.

Henderson: I agree that it is feasible to treat livers with PDT. We treated a mouse liver (7 mm diameter spot) with 5 mg/kg Photofrin II and 113 J/cm^2 from the outside. The damage was about 10 μm deep and the mouse was otherwise completely unscathed, because of the limited light penetration in this tissue (Henderson & Bellnier, this volume). We didn't even see as much necrosis as Dr Bown reported.

Bown: We had the fibres in the middle of the liver.

Henderson: You probably also get a heating effect.

Bown: Thermal damage was accounted for in the control experiments and was much less extensive than the PDT damage.

Dougherty: I wouldn't attempt to treat liver metastases. If individual nodules are found in the liver, surgeons can excise them.

Bown: It is not easy in the middle, but it can be done if they are on the surface. However, interstitial therapy with a laser fibre done percutaneously under ultrasound control is much simpler than surgery. We have started doing this on patients. We use thermal effects from a low power Nd-YAG laser.

Dougherty: That may be one application, but if there are widespread metastases in the liver you can't illuminate the entire liver.

Bown: Even that may be possible if it is done in sections. The liver all regenerates; where you had tumour you may regain normal liver. But it will be some years before we can put this into practice.

MacRobert: The tetra(hydroxyphenyl)porphyrins (THPPs) are not water soluble at neutral pH. How were they dissolved?

Berenbaum: You can dissolve them in either dilute alkali and phosphate-buffered saline (PBS) or dimethyl sulphoxide (DMSO). It doesn't make much difference to the dose–response curve.

MacRobert: Have you tried using an emulsion?

Berenbaum: We can put them in a commercially available lipid emulsion that can be given intravenously.

Dougherty: Have you tried cremophor?

Berenbaum: Not for porphyrins, but we have put β-carotene in cremophor.

Moan: What happens when you dissolve a drug in cremophor or DMSO and dilute that to an aqueous solution for injection?

Kessel: The relatively hydrophobic dyes we studied stay in the cremophor, to my amazement. The fluorescence emission spectrum, which is assumed to be characteristic of the environment, did not change when we diluted the drug in cremophor with water. Uroporphyrin, which is very hydrophilic, might come out of the emulsion.

Moan: That means that the drug–cremophor complex may be carried around in the circulation.

Jori: This is a phase partitioning between the aqueous medium and the cremophor, whereas in DMSO drugs such as phthalocyanines aggregate as the solution is diluted and eventually precipitate.

Dougherty: These water-insoluble compounds tend to be soluble in organic solvents that are water soluble. One can use solvents such as tetrahydrofuran and dilute the solution into aqueous media that contain a detergent such as Tween 80. The dye is transferred into the detergent and the organic solvent can be removed by dialysis. The result is a micellar solution of the drug.

Moan: One must take care when using dyes that are not water soluble because the 'emulsion' changes with time. When you dilute with water or inject the emulsion in animals, the results depend on the timing of the experiment. Conflicting data are produced when different procedures are used.

Dougherty: Yes, the timing is important. Professor Bonnett mentioned the chlorin e_6 aspartic acid ester which is short acting; it has no tendency to be retained *in vivo*. This should be a very amphiphilic material because it has all the charges on one side and hydrophobic groups, such as vinyl, on the opposite side.

Brown: What is the definition of amphiphilic? Are amphiphilic dyes those that are soluble in both water and lipid or those that have polar groups on one side and hydrophobic groups on the other? If they are the latter, the tetra(hydroxyphenyl)porphyrins (structures 9, 10, 11) are not amphiphilic.

Bonnett: Patches of those molecules are polar and patches are non-polar. They have an affinity for both lipid and water. I think that's what amphiphilic normally means.

Dougherty: Have you measured the partition coefficients?

Bonnett: David Kessel (1977) suggested the importance of the partition coefficient. I think there's still more mileage in that idea. In the 2-octanol/PBS system the tetra(hydroxyphenyl)porphyrins are essentially entirely in the lipid phase.

Kessel: These dyes will dissolve in a remarkable variety of solvents. *N*-Aspartyl chlorin e_6 has four carboxyl groups on one side of the molecule and will dissolve in, for example, water or methylene chloride. With the right charge distribution on the molecule, you can use an extremely wide range of solvents. On the other hand, when the carboxyls are evenly spaced, as in uroporphyrin or coproporphyrin, the dye will not dissolve in non-polar organic solvents.

Bonnett: I am keen on the idea of tuning the properties by changing the number of hydroxyl groups. Certainly the solubility properties of some of the compounds are unexpected.

Moan: After your reports on the tetra(hydroxyphenyl)porphyrins, several people made similar phthalocyanines and have run into trouble with water solubility.

Bonnett: The basic systems are really very different. The phthalocyanines have an extra benzene ring on each corner and have a very low affinity for water. For example, copper phthalocyanine is a marvellous stable pigment, but it is not soluble in water at all. It seems that the hydroxyl group does not have enough solubilizing power to overcome the low water solubility of these large aromatic systems. However, the solubility is increased by addition of ionic substituents such as sulphonic acid.

Morgan: In the tetra(hydroxyphenyl)porphyrins the aromatic rings are at 90° to the plane of the porphyrin, but in phthalocyanines they will be flat. Thus the electronic structures are different.

Dougherty: Dr Berenbaum, can you prepare these porphyrin derivatives for injection by dissolving them in sodium hydroxide and neutralizing the solution?

Berenbaum: Some of them won't dissolve in anything, including sodium hydroxide. Some, such as octaethyl β-oxochlorin (17), only dissolve in hot DMSO.

Bonnett: Octaethyl β-oxochlorin (17), which has a chlorin-type chromophore but a low solubility in hydrophilic solvents, is inactive in the *in vivo* bioassay. Problems with delivery may often have been the reason for negative results. Initially we didn't realize that water-insoluble compounds may precipitate before they are able to reach the target site.

Berenbaum: When octaethyl β-oxochlorin is administered i.p. in DMSO no biological effect is observed and solid deposits of drug are found in the liver.

Jori: We found the same. We tried to separate lipoproteins in serum by column chromatography and ultracentrifugation, but the DMSO-injected phthalocyanines and naphthalocyanines were found to have precipitated. That is a problem, because the drug may then be gradually taken up by lipoproteins and transported to tissues in an uncontrolled way.

Dougherty: I suggest that you put the DMSO solution into Tween 80 solution, making a micellar suspension for injection. That should distribute evenly.

Kessel: The National Cancer Institute has a formulation for drugs that are difficult to dissolve. They continue to add carboxymethyl cellulose, ethanol and water and homogenize until they obtain a fine dispersion that can be injected into mice. But occasionally a compound precipitates whatever you do.

Dougherty: What about the chlorin series versus the chlorophyll series? Chlorin e_6, is not a particularly good photosensitizer; it's a short-acting material, as is the aspartic acid derivative. But the phytyl group introduces enough hydrophobicity for chlorophyll to be a perfectly acceptable *in vivo* photosensitizer. Chlorophyll dissolves easily in Tween 80.

Jori: I have not worked with chlorins *in vivo*. But I recall early photosensitization studies with pheophytin and chlorophyll. Chlorins are very rapidly photo-oxidized. Is this also true for the bacteriochlorins and chlorins that are used *in vivo*?

Dougherty: Yes, but whereas bacteriochlorophyll is extremely sensitive to light and to oxidation *in vitro*, it appears to be much more stable *in vivo*. For example, in trying to measure singlet oxygen yields with it, I tried to trap with tryptophan or diphenyl isobenzofuran, but even at high concentrations I could not prevent self-oxidation. One might therefore predict that it would be impossible to use this material in animals, but *in vivo* it lasts considerably longer.

Jori: It must be stabilized by some environmental factor.

Brown: How efficient is chlorophyll itself as a photosensitizer?

Dougherty: It is not particularly efficient, in the sense that the doses required are higher than those anticipated from the increased red absorption. However, 10 mg/kg, activated after 24 h at 675 nm with light doses of 150 J/cm^2, gives good tumour control in animals. This effect is of the same order as that obtained with Photofrin II. Chlorophyll is clearly not as stable, but it will persist for a few days. We evaluate normal tissue damage using the foot reaction, not a skin reaction. Three or four days after injection we still see reactions, but they cease after four or five days. Thus chlorophyll slowly clears; perhaps it's just oxidized.

Jori: Photoexcited phenols are usually type I photosensitizers. They easily undergo hydrogen transfer with suitable substrates. Does something similar occur with the hydroxyl-substituted chlorins or porphyrins?

Bonnett: That is possible. We haven't examined chlorins, but we tried to photo-oxidize the porphyrins *in vitro*. I was expecting phenolic oxidation and perhaps coupling. I thought we might obtain another sort of dimer system, which would have been a nuisance. However, none of the three isomers (*ortho*, *meta*, *para*) of THPP photo-oxidizes at an appreciable rate in neutral organic solvents. If you make the solutions alkaline they do begin to decompose, but slowly.

Dougherty: Photofrin II does photo-oxidize. What is known about the photo-oxidation products?

Moan: I think the aggregates are quite stable. The monomers and the dimers decompose by a scheme similar to that known for haematoporphyrin.

Dougherty: What about in cells?

Moan: The photodegradation of Photofrin II is much faster in cells than in aqueous solution. In cells and serum, reactions with proteins are involved. Our results agree with those of Krieg and Whitten (see Moan et al 1988 and references therein).

Professor Bonnett lists some requirements for a good drug. In view of our discussion, we should also emphasize the properties of the drug in solution. We have been working with drug–serum mixtures, measuring the aggregated and monomeric fractions as well as the amounts bound to the different serum proteins. The pattern changes with time, which means that for clinical purposes the drug solutions should be prepared and used in a reproducible manner. It would be advantageous to have drugs that allow some variation in the handling procedure.

Bonnett: I agree. Porphyrin chemists are accustomed to seeing water-containing solutions that are initially clear become opalescent and then form a precipitate. We must avoid that. One way to do so is to enhance hydrogen bonding to solvent water molecules by using polar substituents, such as hydroxyl groups. That should reduce aggregation of the compound, but I haven't done the experiment yet.

Moan: In the absence of serum *m*-THPP aggregates readily. The only dyes we tested that do not aggregate are tetraphenylporphyrin tetrasulphonate and, apparently, aluminium phthalocyanine tetrasulphonate.

Dougherty: Neither of which is a good *in vivo* photosensitizer.

Brown: But this must depend on dose. The lower the dose you can use the less chance there is of aggregation in serum.

Moan: *m*-THPP is very lipophilic, and has a strong tendency to aggregate in water.

References

Bown SG, Tralau CJ, Coleridge-Smith PD, Akdemir D, Wieman TJ 1986 Photodynamic therapy with porphyrin and phthalocyanine sensitisation: quantitative studies in normal rat liver. Br J Cancer 54:43–52

Henderson BW, Bellnier DA 1989 Tissue localization of photosensitizers and the mechanism of photodynamic tissue destruction. In: Photosensitizing compounds: their chemistry, biology and clinical use. Wiley, Chichester (Ciba Found Symp 146) p 112–130

Kessel D 1977 Effects of photoactivated porphyrins at the cell surface of leukemia L1210 cells. Biochemistry 16:3443–3449

Moan J, Rimington C, Malik Z 1988 Photoinduced degradation and modification of Photofrin II in cells in vitro. Photochem Photobiol 47:363–367

van Lier JE, Spikes JD 1989 The chemistry, photophysics and photosensitizing properties of phthalocyanines. In: Photosensitizing compounds: their chemistry, biology and clinical use. Wiley, Chichester (Ciba Found Symp 146) p 17–32

Photodynamic therapy: light delivery and dosage for second-generation photosensitizers

Brian C. Wilson

Hamilton Regional Cancer Centre and McMaster University, 711 Concession Street, Hamilton, Ontario, Canada L8V 1C3

Abstract. With the development of new photosensitizers that have enhanced photoactivation at longer wavelengths than haematoporphyrin derivative, new considerations arise in the light source and delivery systems and in the techniques for physical dosimetry and *in vivo* optical measurements in photodynamic therapy. The limitations and future potential of solid-state laser sources are presented. The relationships between photosensitizer photoactivation characteristics and the effective photodynamic treatment volume are developed and discussed quantitatively. The problems in defining and measuring the photodynamic dose are examined, and potential techniques for measuring the factors involved in this are evaluated with emphasis on noninvasive approaches which may be used clinically.

1989 Photosensitizing Compounds: their Chemistry, Biology and Clinical Use. Wiley, Chichester (Ciba Foundation Symposium 146) p 60–77

Clinical photodynamic therapy (PDT) is currently based on the photosensitizer haematoporphyrin derivative (HpD), or variants of this compound. Despite the limitations of HpD, clinical efficacy has been demonstrated for a range of solid tumours (Dougherty 1986). There is growing interest in the next generation of clinical photodynamic agents which may overcome the biophysical, biochemical and biological shortcomings of HpD. Among the goals are improved tumour-to-normal tissue selectivity, greater photoactivation efficiency and activation at longer wavelengths. It is expected that these developments will allow successful treatment of larger tumours than are treatable with HpD at 630 nm, as a result of increased effective penetration of the photoactivating light into the tissue.

This paper presents some of the technological and biophysical factors associated with photosensitizers activated in the far-red/near-infrared regions, between 650 and 850 nm. Possible new laser sources are reviewed, with emphasis on semiconductor diode lasers and solid-state tunable lasers. The possible increase in tumour volume treatable with longer wavelength photosensitizers will be evaluated by a quantitative analysis of light penetration in tissues containing these photosensitizers. Finally, issues involved in defining and measuring the major biophysical factors, such as light fluence, photosensitizer concentration and tissue oxygenation, which contribute to the 'photodynamic dose' are discussed.

Laser sources for PDT

Although conventional incandescent or arc lamps with wavelength filtering were used in early clinical studies of PDT, and still have useful roles as illumination sources in the laboratory and for treating surface lesions, lasers have become the standard light source for most PDT applications, largely because high light power can be coupled into single-strand optical fibres for endoscopic or interstitial use. Because HpD is activated at around 630 nm, the energy density delivered to the tumour may be up to several hundred J/cm^2 for surface treatments. Thus, for reasonable treatment times, source powers of up to several Watts may be required. It is likely that second-generation photosensitizers will have similar power requirements in order that larger lesions may be treated.

Details of current and potential lasers for clinical PDT are summarized in Table 1. The first three have already been used clinically with HpD, the dye laser pumped by a 5–25 W argon ion laser being the most common system. The dye laser efficiency is up to about 20% under optimal operating conditions, so that several Watts of red light can be obtained with a high power argon-pumped laser, of which 80–90% can be coupled into optical fibres (200–600 μm core diameter). The main disadvantages of the argon-pumped dye laser are the high capital and running costs, poor reliability in the clinical environment, large size and immobility. The flow-cell dye laser pumped by a copper metal vapour laser is an alternative in which comparable powers can be achieved. The gold vapour laser is simpler and cheaper, but has the limitation of operating at a single fixed wavelength suitable only for HpD. It also has a very large beam size, making coupling into optical fibres inefficient.

Although these dye lasers are suitable for any photosensitizer by tuning across the dye emission spectrum or changing the dye for a different wavelength range, their present complexity and cost are limiting factors in setting up PDT programmes. Thus there is considerable interest in the development of alternative laser sources, both for HpD and for longer wavelength photosensitizers. There are two main possibilities: semiconductor diode lasers and tunable solid-state lasers.

Diode lasers

In the 790–830 nm wavelength range, commercially available diode lasers can give up to about 1 W, either continuous wave (CW) or electronically pulsed.

TABLE 1 Characteristics of lasers for photodynamic therapy[a]

Laser type	*Wavelength (nm)*	*Power output*[b]	*Approximate price ($\$ \times 10^{-3}$)*	*Comments*
Argon ion-pumped dye	<900 tunable	CW to several watts	100+	dye jet
Copper vapour-pumped dye	<900 tunable	mJ, 10 ns, kHz, average to several watts	100	dye flow-cell
Gold vapour	628 fixed	mJ, 10 ns, kHz, average to ~10 watts	50+	>20 mm diameter beam
Diode (GaAlAs)	790–830 fixed, selectable	CW, P ~1 watt	5+ per watt	50–60% coupling into single-strand fibre
Alexandrite (flashlamp-pumped)	720–780 (701–826 theoretical) tunable	0.1–1 J, 100–400 μs, 10 Hz average to several watts	50+	
Titanium sapphire argon-pumped (20 W)	660–990	CW >1 W 690–980; >2 W 720–950; >3 W 750–900	80+	
[Frequency-doubled YAG-pumped, enhanced flashlamp-pumped]		[P]		
Pulsed dye (frequency-doubled YAG, excimer, or flashlamp-pumped)	<900 tunable	P		dye jet

[a]Technical details after Hecht (1986).
[b]CW, continuous wave; P, pulsed.

The beam quality makes for rather poor coupling into small core fibres (50–60%). A total single-fibre output power exceeding 1 W can be obtained by polarization coupling of two separate 1 W diode lasers at the fibre input, at a total cost of around $10 000 (diodes + simple power supplies). This output power is likely to double in the near future with single-element, 2 W diode lasers. In addition, several watts could be obtained in a small multiple-fibre bundle. Other very high power diode lasers (about 5 W) are diode arrays, but efficient coupling into single optical fibres is not presently available.

Diode lasers are not tunable within the quoted wavelength range, but the wavelength may be pre-selected at some additional cost. There is a divergence of opinion on whether high power (> 1 W), single-element diode lasers which operate down to 700 nm or below will be available in the near future. Much will depend on the development of suitable semiconductor materials and on the market for large volume applications, such as laser disks. The potential advantages of diode lasers for clinical PDT are the low capital cost, negligible running costs, high reliability, and small size and portability. Each of these would have a major impact on the clinical availability and flexibility of PDT.

Vibronic solid-state lasers

These lasers may be considered as solid-state analogues of dye lasers. Their crystalline structure allows excitation of vibrational as well as electronic energy levels, leading to an emission continuum which permits wavelength tuning over a wide range. Like the dye laser, these are passive devices, and need to be pumped by either a CW source, such as an argon ion laser, or a pulsed source, such as a flashlamp or frequency-doubled YAG laser. The flashlamp-pumped Alexandrite laser and the argon ion-pumped titanium sapphire laser are commercially available with average powers of several watts. In both cases the practical wavelength range giving high laser power is much more restricted than the theoretical range. Nevertheless, both are feasible lasers for photosensitizers in the near-infrared, and there is the potential for reaching below 700 nm with the titanium sapphire laser in the future.

The flexibility of simple wavelength tuning is offset presently by the higher cost, greater complexity and reduced portability compared with diode lasers. However, the vibronic lasers should be substantially more reliable than the corresponding dye lasers.

Other lasers

As shown in Table 1, there are a variety of pulsed dye lasers with different pump sources which may find a role in clinical PDT and other medical applications. If there should prove to be significant advantages to specific pulsed irradiation techniques, these lasers may provide a useful, selectable range of power, pulse and wavelength characteristics. However, they are likely to have the same problems of complexity and size as the established dye laser systems.

Effective treatment volume with longer wavelength photosensitizers

A major motivation for developing new photosensitizers is the possibility of treating larger tumour volumes because of the greater tissue penetration of longer wavelength light. Ignoring the subsurface build of fluence, the fluence in a tissue irradiated by a wide incident collimated beam decreases with depth, d, approximately as $e^{-d/\delta_{eff}}$ where δ_{eff} is the effective penetration depth (Jacques 1989, Profio & Doiron 1987, Wilson & Patterson 1986). The effective penetration depth is inversely proportional to the effective attenuation coefficient, which has contributions from both optical absorption and optical scatter of the tissue. Figure 1 shows representative curves for the wavelength dependence of δ_{eff} in

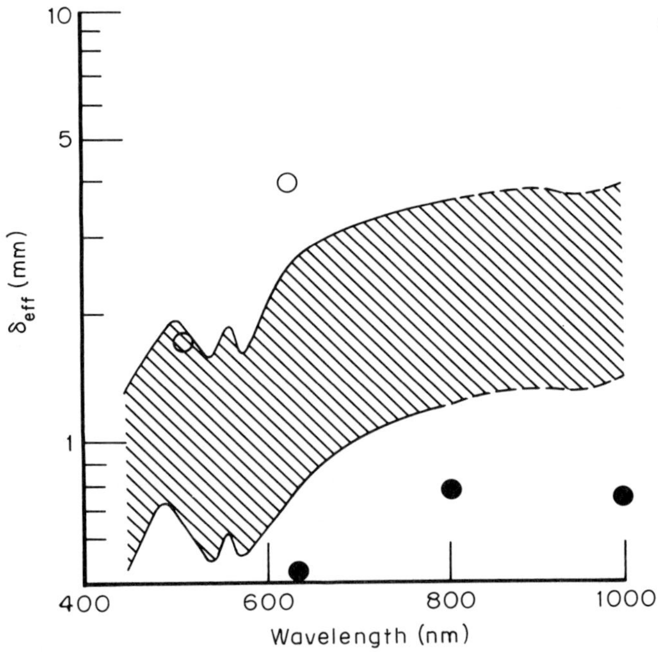

FIG. 1. Schematic diagram of the wavelength dependence of effective penetration depth in soft tissues. The curves are based on the data of Wilson et al (1985), with extrapolation above 800 nm from those of Bolin et al (1984). The upper range corresponds to lightly-pigmented tissues (e.g. brain, white muscle), and the lower range to highly-pigmented tissues (e.g. liver). The structures at 500–600 nm and 900–1000 nm are oxyhaemoglobin and water absorption peaks, respectively. The individual data points represent extremes in the published results: ○, neonatal brain *in vitro* (Svaasand & Ellingsen 1983); ●, rat liver *in vitro* (Parsa et al 1989).

soft tissue. The penetration depth increases from typically 1–2 mm at 630 nm to about twice this at 700–850 nm, although, as indicated, there are particular tissues with considerably higher or lower penetration. This is probably mainly because of reduced absorption, particularly of haemoglobin, but the scatter coefficient may also decrease somewhat.

A number of studies (Bown et al 1986, van Gemert et al 1985) have shown a well-demarcated zone of tissue necrosis at some effective treatment depth, d_t, following PDT, rather than a gradual transition from full necrosis at the higher light levels to no damage at deeper levels in the tissue. The probable explanation is that a minimum threshold level of the photo-generated cytotoxic agent, such as singlet oxygen, is required to produce necrosis. In the simplest model (Profio & Doiron 1981), the production of singlet oxygen is proportional to the product of the light fluence and the absorption coefficient, μ_{ap}, of the photosensitizer, and also depends on the tissue oxygenation. In turn, $\mu_{ap} = \epsilon_p.C$, where ϵ_p is the photosensitizer extinction coefficient at the treatment wavelength and C is the photosensitizer concentration in the tissue. For haematoporphyrin derivative, ϵ_p at 630 nm is approximately $7 \times 10^{-3}\,mm^{-1}$ $(\mu g/g)^{-1}$. At clinically administered doses of 1–5 mg/kg body weight, C in tumour tissue is of the order of μg/g, and adequate light irradiation appears then to give an effective treatment depth in solid tumours of about 5–10 mm, i.e. about 3–5 effective penetration depths.

The effective treatment depth (or volume) may be increased by altering both the biological and physical characteristics of the photosensitizers used. Here we shall consider changing only the physical characteristics, for example increasing μ_{ap} by increasing ϵ_p and/or C, and increasing δ_{eff} of the tissue by using a longer activation wavelength. Figure 2A shows d_t as a function of δ_{eff} for surface irradiation, over a range of values of the ratio $R = \mu_{ap}/\mu_{HpD\text{-}630}$, where $\mu_{HpD\text{-}630}$ is the absorption coefficient of HpD at 630 nm for a tissue concentration of 2 μg/g. These curves are for fixed incident light fluence, i.e. for the same value of the product of incident power density and treatment time, using an assumed 'standard' HpD treatment of 10 mm effective treatment depth in tissue with an effective tissue penetration depth of 2 mm. It is seen that d_t is proportional to the effective penetration depth, with the gradient increasing with the photosensitizer ratio, R. However, for a given tissue, d_t increases only incrementally with increasing R. Thus, for example, if the concentration of the photosensitizer is the same as that of HpD but the extinction coefficient is 20 times higher, the effective treatment depth will increase by only 3 penetration depths ($e^3 \approx 20$).

From Fig. 2A, it might be concluded that effective treatment depths can be achieved which are several times greater than those with HpD. However, there are factors not included in these simple curves which may markedly alter this conclusion, in particular the reduction in the effective penetration depth due to the light absorption by the photosensitizer. As μ_{ap} increases, so also does

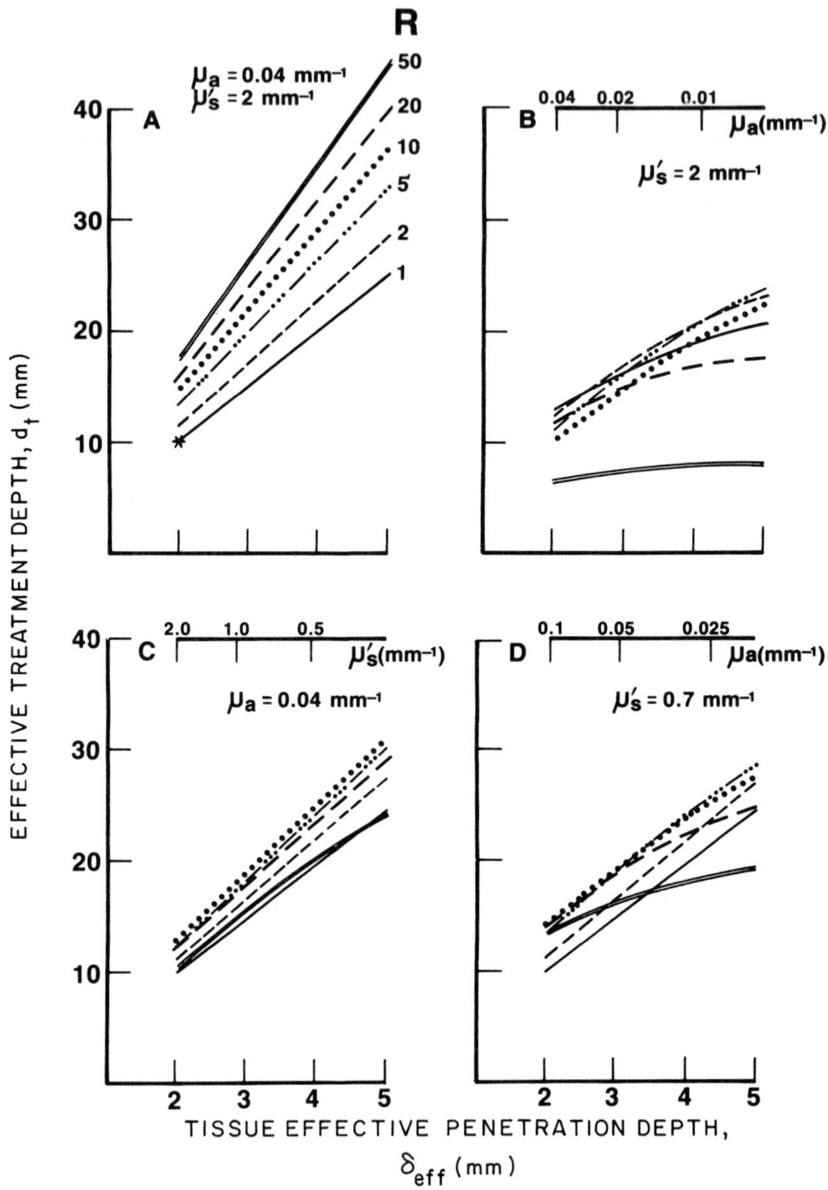

FIG. 2 *(see legend on opposite page)*.

the total optical attenuation of the tissue. The magnitude of this 'self-shielding' effect depends on μ_{ap} and on the inherent optical absorption and scattering of the tissue. In previous studies (Wilson et al 1986, Wilson & Patterson 1989), we showed that for HpD at 630 nm the reduction in the effective penetration depth by self-shielding was less than a few percent, even for very lightly-pigmented tissues. By estimating the reduction in penetration depth by self-shielding for photosensitizers with a range of absorptions and for a variety of tissue types we found that it could be significant. However, we did not consider the effect of the reduced tissue absorption coefficient (or scatter coefficient) in using longer wavelength photosensitizers. We have, therefore, modelled this more realistic situation using diffusion theory for a number of different conditions, and the results are shown in Figs 2B–D. (As in Wilson et al (1986), the relationship $\delta^2_{eff} = 1/[3(\mu_a + \mu_{ap})(\mu_a + \mu_{ap} + \mu'_s)]$ was used, where μ_a and μ'_s are the inherent absorption and transport scattering coefficients of the tissue at the photosensitizer activation wavelength.)

Figures 2B and 2C are for tissue with low inherent absorption. In Fig. 2B, the increase in effective penetration depth of the tissue (corresponding to longer wavelength) has been ascribed entirely to decrease in the tissue absorption coefficient. In this case: (1) The increase in d_t is substantially suppressed, reaching only about twice that of the standard treatment; (2) There is an optimum value of R, beyond which d_t actually decreases due to the photosensitizer self-shielding; (3) Within a wide range the treatment depth is insensitive to R; and (4) The reduction in d_t due to photosensitizer self-shielding is greatest for high values of R, as illustrated in Fig. 3.

Figure 2C shows the other extreme where the tissue absorption is fixed and all the increase in δ_{eff} is due to reduced scattering. Even in this unlikely case, the same general conclusions apply, although the optimum R value is higher and the suppression of d_t is less severe. If the tissue has more blood or other pigments, so that its absorption is higher, then the magnitude of the effect is also reduced, as illustrated in Fig. 2D, but the overall behaviour is the same. The conclusions also hold if the standard conditions are altered, for example if d_t for HpD at 630 nm is a different number of penetration depths for a particular tissue.

FIG. 2. Effective treatment depth versus inherent optical penetration depth of tissue for a range of photosensitizer ratios and four assumed conditions. A. Assuming no photosensitizer self-shielding. * represents the standard treatment conditions for this model. B. Assuming that increase in δ_{eff} of the tissue is due to decrease in tissue absorption (μ_a) from 0.04 mm^{-1} at $\delta_{eff} = 2$ mm, with tissue scattering (μ'_s) constant at 2.0 mm^{-1}. Photosensitizer self-shielding included. C. Assuming that increase in δ_{eff} of the tissue is due to decrease in μ'_s from 2.0 mm^{-1} at $\delta_{eff} = 2$ mm, with μ_a constant at 0.04 mm^{-1}. Photosensitizer self-shielding included. D. As for B, but for more highly pigmented tissue ($\mu_a = 0.1$ mm^{-1}, $\mu'_s = 0.7$ mm^{-1} at $\delta_{eff} = 2$ mm).

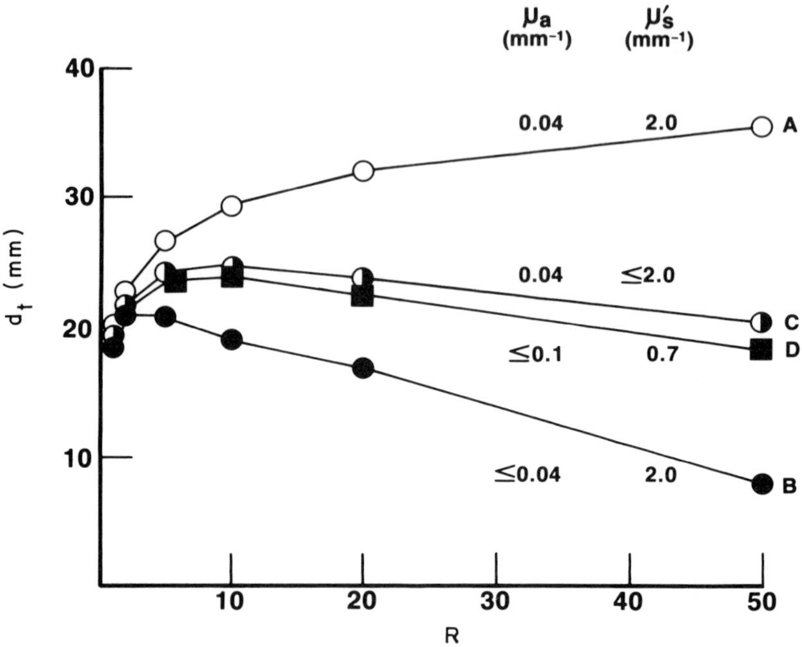

FIG. 3. Effective treatment depth versus photosensitizer ratio (R) for $\delta_{eff} = 4$ mm. The curves A–D correspond to those of Fig. 2.

The implications of these findings for the use of new photosensitizers are: firstly, that the extent to which the effective tumour volume can be increased depends in a complex way on both the characteristics of the photosensitizer (extinction coefficient, tissue uptake) and on the optical properties of the target tissue at the activation wavelength; and secondly, that the optimum conditions do not necessarily correspond to maximum photosensitizer concentration and extinction, and establishing these conditions requires quantitative knowledge of the optical properties of the tissue involved. Finally, additional studies are required to assess effects such as photosensitizer photobleaching or saturation and light-induced tissue heating. For example, photobleaching could possibly either increase or decrease the effective treatment depth: it causes a reduction in the amount of singlet oxygen produced per unit light fluence at depth, but counteracting this is the reduced optical shielding as the photosensitizer is destroyed. Similarly, with pulsed laser irradiation transient photosensitizer saturation in the surface tissues exposed to very high power density may reduce the optical shielding, allowing greater light penetration. Further studies will also be needed to determine if the increased light penetration at longer wavelengths will allow a greater incident power density without unacceptable tissue heating. If so, and if suitable higher power laser sources are available, then the treatment times could be reduced, or a greater effective treatment depth achieved for the same treatment time. However, as with increasing the photosensitizer absorption, the increase in d_t will only be incremental: tripling the incident fluence by tripling the fluence rate or the treatment time only adds one penetration depth to d_t.

Measurement of the photodynamic dose

For PDT using new photosensitizers, the establishment of a meaningful 'photodynamic dose', and the development of methods to measure this will be important (Potter 1989, Profio & Doiron 1981, 1987, Wilson & Patterson 1986). As illustrated in Fig. 4, the present PDT treatment prescription is simply in terms of the incident light fluence and the administered quantity of photosensitizer. This does not account for the differences in light penetration in individual tumours, only partially accounts for differences due to irradiation geometry, and ignores the variation in tumour uptake of the photosensitizer. These uncontrolled factors presumably contribute to the heterogeneity in tumour response.

The next level of photodynamic dose specification will include the depth distribution of the light, the concentration of photosensitizer in the tumour, and the tumour oxygenation. Further refinements (Fig. 4, level 3) may be to generate iso-fluence, photosensitizer concentration and tissue oxygen

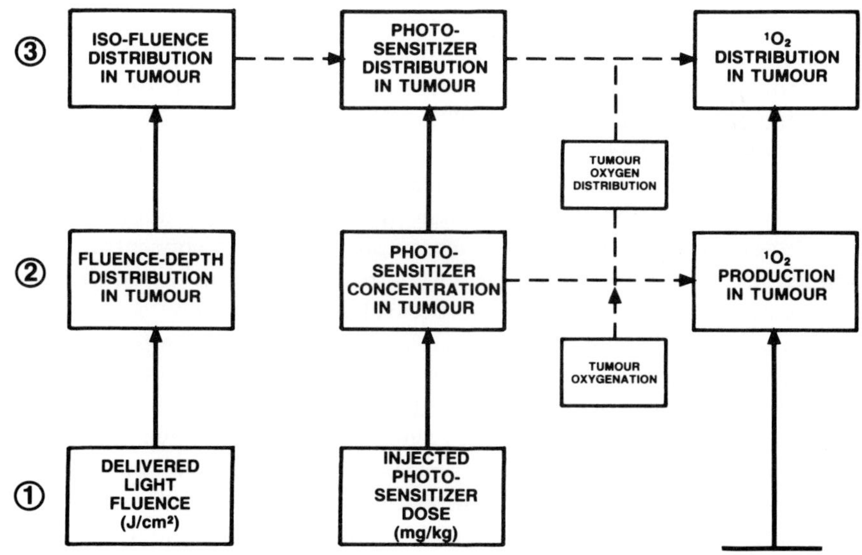

FIG. 4. Schematic of the potential levels of sophistication in defining and measuring the photodynamic dose in tumour tissue.

distributions. An alternative approach is to determine directly the singlet oxygen production or distribution. Here we shall examine briefly the status of individualized light dose, photosensitizer concentration and singlet oxygen measurements, and consider how these may be affected by using longer wavelength photosensitizers.

Light dosimetry

The determination of light distributions in tissue may be either by direct spot measurements of the local fluence rate using optical fibre light detectors (Marijnissen & Star 1987), or by determining the optical absorption and scattering properties of the tissue and calculating the fluence distribution using a radiation transport model (Star et al 1987). A number of techniques are under development to allow determination of tissue optical properties noninvasively *in vivo* (Jacques 1989, Patterson et al 1989a, Wilson et al 1989a,b) and enable more detailed estimation of light dose distributions in individual patients. In either case, use of longer wavelengths will probably be advantageous, because the influence of haemoglobin absorption will be less than at 630 nm, so that variations in tissue blood content and oxygenation will have less effect on the light distributions.

Photosensitizer concentration

Optical methods to measure photosensitizer concentration in tissue are based on quantitative fluorometry (Profio & Sarnaik 1984) or reflectance spectrophotometry (Patterson et al 1987, 1989b). The usefulness of fluorometry for new photosensitizers will depend on their fluorescence characteristics. Excitation of fluorescence at the PDT treatment wavelength, as is possible with HpD (630 nm excitation, 690 nm emission), is particularly valuable (Potter & Mang 1984) because it allows the average photosensitizer concentration to be 'sampled' over the effective treatment volume, rather than just at the surface of the tumour as in the case of short-wavelength activation.

Reflectance spectrophotometry, i.e. the quantitation of the photosensitizer absorption 'signature' in the spectrum of light diffusely reflected from the tissue, is suitable for any photosensitizer regardless of its fluorescence characteristics, and inherently samples the treatment volume. It is substantially better at longer wavelengths with highly absorbing photosensitizers, because the signature is stronger and the effect of blood absorption is reduced. Both fluorometry and reflectance spectrophotometry require a knowledge of the tissue optical properties for calculation of absolute values of the photosensitizer concentration. Therefore these techniques will also benefit from noninvasive optical dosimetry (Patterson et al 1989b).

Finally, we note that photosensitizer radiolabelling with γ ray or positron-emitting radionuclides suitable for quantitative nuclear scanning may be possible with some of the new metal-containing photosensitizers (Wilson & van Lier 1989).

Oxygenation

As indicated in Fig. 4, the photodynamic dose prescription should include the tumour oxygenation together with measurement of light fluence and photosensitizer concentration. Optical techniques may enable oxygen measurements to be made noninvasively, based on reflectance or transmittance spectrophotometry, as demonstrated, for example, in cerebral monitoring (Delpy et al 1988). The extent to which the local oxygenation in tumour tissue can be determined by time-resolved optical spectroscopy is under investigation (B. Chance, personal communication).

Singlet oxygen

Direct noninvasive measurement of the singlet oxygen generated in tissue is possible in principle by measuring the 1270 nm luminescence emission of singlet oxygen de-excitation. Observation of this has been reported *in vivo* (Parker 1987) using an optical chopping technique to remove the background infrared fluorescence from the tissue and the photosensitizer. However, to date, other groups have not been able to confirm this, and it is possible that the lifetime of singlet oxygen *in vivo* is too short to allow reliable detection by this technique with available infrared detectors. It may still be feasible to measure the 1270 nm emission by using a pulsed laser technique where the pulse length is much shorter than the lifetime of the photosensitizer triplet state and detecting in single photon counting mode (M. S. Patterson, personal communication).

A possible alternative would be to detect optically the changes in a chemical marker in the tissue which is affected by singlet oxygen, such as, for example, the destruction of a fluorescent dye (Profio 1989). However, this complicates the problem, because then the uptake and distribution of the marker is important and needs to be controlled or measured.

Conclusions

In summary, the clinical introduction of second-generation photosensitizers provides opportunities for the use of new laser technologies. Specific biophysical factors, such as the effective treatment depth, will be more complex to control and will require more sophisticated and detailed measurements of light penetration and photosensitizer concentration in individual patients than at present. However, as the optically based methods for these measurements become easier, more reliable and more accurate, so the improved control of photodynamic treatments will be possible.

Acknowledgement

This work is supported by the National Cancer Institute of Canada.

References

Bolin FP, Preuss LE, Cain BW 1984 A comparison of spectral transmittance for several mammalian tissues: effects at PRT frequencies. In: Doiron DR, Gomer CJ (eds) Porphyrin localization and treatment of tumors. Alan R. Liss, New York, p 211–225

Bown SG, Tralau CJ, Coleridge-Smith PD, Akdemir D, Wieman TJ 1986 Photodynamic therapy with porphyrin and phthalocyanine sensitization—quantitative studies in normal rat liver. Br J Cancer 54:43–52

Delpy DT, Cope M, van der Zee P, Arridge S, Wray S, Wyatt J 1988 Estimation of optical pathlength through tissue from direct time of flight measurement. Phys Med Biol 33:1433–1442

Dougherty TJ 1986 Photosensitization of malignant tumors. Semin Surg Oncol 2:24–27

van Gemert JC, Berenbaum MC, Gijsberg GHM 1985 Wavelength and light-dose dependence in tumour phototherapy with hematoporphyrin derivative. Br J Cancer 52:43–49

Hecht J 1986 The laser guidebook. McGraw-Hill, New York

Jacques SL 1989 Simple theory, measurements and rules of thumb for dosimetry during photodynamic therapy. Proc International Society for Optical Engineering (SPIE) 1065:100–108

Marijnissen JPA, Star WM 1987 Quantitative light dosimetry in vitro and in vivo. Lasers Med Sci 2:235–242

Parker J 1987 Optical monitoring of singlet oxygen generation during photodynamic treatment of tumors. IEEE (Inst Electr Electron Eng) Circ Dev, 10–21

Parsa P, Jacques SL, Nishioka NS 1989 Optical properties of rat liver from 350 to 2200 nanometers. Appl Opt 28:2325–2330

Patterson MS, Wilson BC, Feather JW, Burns DM, Pushka W 1987 The measurement of dihematoporphyrin ether concentration by reflectance spectrophotometry. Photochem Photobiol 46:337–343

Patterson MS, Chance B, Wilson BC 1989a Time resolved reflectance and transmittance for the noninvasive measurement of tissue optical properties. Appl Opt 28:2331–2336

Patterson MS, Schwartz E, Wilson BC 1989b Quantitative reflectance spectrophotometry for the noninvasive measurement of photosensitizer concentration in tissue during photodynamic therapy. Proc International Society for Optical Engineering (SPIE) 1065:115–122

Potter WR 1989 PDT dosimetry and response. Proc International Society for Optical Engineering (SPIE) 1065:88–99

Potter WR, Mang TS 1984 Photofrin II levels by in vivo fluorescence photometry. In: Doiron DR, Gomer CJ (eds) Porphyrin localization and treatment of tumors. Alan R. Liss, New York, p 177–186

Profio AE 1989 Measurements of singlet oxygen in photodynamic therapy. Proc International Society for Optical Engineering (SPIE) 1065:in press

Profio AE, Doiron DR 1981 Dosimetry considerations in phototherapy. Med Phys 8:190–196

Profio AE, Doiron DR 1987 Transport of light in tissue in photodynamic therapy. Photochem Photobiol 46:591–599
Profio AE, Sarnaik J 1984 Fluorescence of HPD for tumor detection in photoradiation therapy. In: Doiron DR, Gomer CJ (eds) Porphyrin localization and treatment of tumors. Alan R. Liss, New York, p 163–165
Star WM, Marijnissen JPA, van Gemert MJC 1987 Light dosimetry: status and prospects. J Photochem Photobiol 1:149–168
Svaasand LO, Ellingson R 1983 Optical properties of human brain. Photochem Photobiol 38:283–299
Wilson BC, van Lier JE 1989 Radiolabelled photosensitizers for tumor imaging and photodynamic therapy. J Photochem Photobiol 3:459–463
Wilson BC, Patterson MS 1986 Physics of photodynamic therapy. Phys Med Biol 31:327–360
Wilson BC, Patterson MS 1989 The determination of light fluence distributions in photodynamic therapy. In: Kessel D (ed) Photodynamic therapy of neoplastic disease. CRC Press, Boca Raton, in press
Wilson BC, Jeeves WP, Lowe DM 1985 In vivo and post mortem measurements of the attenuation spectra of light in mammalian tissues. Photochem Photobiol 42:153–162
Wilson BC, Patterson MS, Burns DM 1986 Effect of photosensitizer concentration in tissue on the penetration depth of photoactivating light. Lasers Med Sci 1:235–244
Wilson BC, Park Y, Hefetz Y, Patterson M, Madsen S, Jacques S 1989a. The potential of time-resolved reflectance measurements for the noninvasive determination of tissue optical properties. Proc International Society for Optical Engineering (SPIE) 1064:in press
Wilson BC, Patterson MS, Flock ST, Wyman DR 1989b Tissue optical properties in relation to light propagation models and in vivo dosimetry. In: Chance B (ed) Photon migration in tissue. de Gruyter, Berlin, in press

DISCUSSION

Dougherty: Dr Wilson, you show that the use of sensitizers that absorb at longer wavelengths does not achieve the expected increased tissue penetration, because of self-absorption by the sensitizer. Presumably that could be overcome by finding sensitizers that are more efficient generators of singlet oxygen and could, therefore, be used at lower doses. For example, if a dye is ten times more efficient *in vivo*, you could use one-tenth of the amount.

Wilson: Yes. However, the increase in effective treatment depth as you change the photoefficiency or the ratio of the absorption coefficient of the new sensitizer to that of HpD at 630 nm is only an additive effect, even in the absence of self-absorption. For example, phthalocyanines have about 20 times the extinction coefficient of HpD. If you use the same concentration you only add three penetrations depths ($e^3 = 20$), as seen in Fig. 2. It's difficult to markedly increase the treatment depth unless you get onto the slope of the curve. Similarly, a tenfold increase in photoefficiency will change the treatment depth by only two penetration depths.

Dougherty: Two penetration depths is a significant improvement.

Wilson: It's only 4 mm.

Dougherty: It depends on the drug; for one that absorbs at 800 nm it is twice that.

Wilson: Incrementally, it's not going to make much difference. You are presumably already achieving four or five penetration depths; you are not going to double the treatment depth.

Morgan: We are assuming that therapy is being limited by light penetration. The surviving hypoxic fraction in tumours probably contributes more to tumour regrowth than lack of light penetration. When we implant a tumour over a liver and give phototherapy with HpD, we get liver damage but we can't cure the tumour. The light has penetrated through the tumour; that's not the problem. Some other factor is involved, but we don't know whether it is hypoxia. Studying that might be more important than finding drugs which are activated at 800 nm.

We are discussing using 800 nm light, but there may not be much difference between treatments at 700 and 800 nm. Solid-state lasers, which are admittedly more expensive than the diode lasers, are similar to the YAG lasers that are already used clinically. Therefore physicians will be happy to use them. Perhaps the critical wavelength is 690 nm or above, rather than 800 nm.

The naturally occurring bacteriochlorins absorb strongly between 770 and 800 nm whereas the synthetic bacteriochlorin described by Bonnett & Berenbaum (this volume) absorbs maximally at 735 nm. That seems to be about the limit for stable porphyrin derivatives. We've made some that absorb maximally at 765 nm, but they are oxygen-sensitive and decompose immediately. This wavelength limitation is not a problem if the ruby, titanium sapphire or alexandrite lasers are viable options.

Dougherty: Argon-pumped dye lasers can produce the required wavelengths, but diode lasers cannot.

Truscott: Diode lasers produce light at 790–830 nm. The longest wavelength maximum absorption found for naturally occurring bacteriochlorins is 770 nm for bacteriochlorophyll, a full 20 nm shorter.

Dougherty: The *in vivo* action spectrum peaks at 780 nm.

Truscott: It depends on the conditions. We extracted bacteriochlorophyll from carotenoid-less mutant bacteria, and in many solvents or in a liposomal preparation absorption peaks at 770 nm, although in pyridine the peak is near 780 nm. By 790 nm, the absorption has fallen off substantially.

Dougherty: You can activate bacteriochlorophyll *in vivo* at 780 nm without any problem; that is the optimum, which may shift in different conditions. Preparations of bacteriochlorophyll in Tween 80 absorb maximally at 780 nm.

Baer: I have some recent information on these lasers. The titanium sapphire laser can be tuned for continuous wave operation from 653 nm to beyond 1100 nm. By pumping with 25 W we obtain over 3 W in the 670–1000 nm range. At the peak, between 750 and 800 nm, we can obtain over 5 W. In aluminium-doped gallium arsenide lasers you get shorter wavelengths the more aluminium you put in. The limit is about 780 nm at room temperature. By cooling you should be able to go further to the blue. The upper limit, reached by taking all the aluminium out, is currently about 850 nm. With other technologies, such as using other doping materials and stressing the lattice, you can obtain wavelengths beyond 900 nm. At present the limit for commercially available lasers at room temperature is 780 to 850 nm.

Wilson: Would you like to comment on the possibility of obtaining high power at short wavelengths? Someone from Spectra Diode told me that was unlikely.

Baer: I am optimistic!

Dougherty: The only people that I know who are studying this are the Japanese. Is Spectra-Physics researching in that area?

Baer: Not to my knowledge.

Morgan: What are the lifetimes of these diode lasers?

Baer: That depends on many factors, but the current generation of 0.5 W laser diodes have projected lifetimes of 5000 h or more. The 1 W lasers have shorter lifetimes, but still in the order of several thousand hours. The projected lifetimes for lasers using the bar technology, which gives 5 W, are more than 5000 h.

Bonnett: How much power is needed in clinical applications?

Dougherty: One watt is sufficient for most of them, assuming that the new photosensitizers require as much power as dihaematoporphyrin ether.

Bown: It depends what area you want to illuminate.

Dougherty: Yes; for the bladder and a large area of skin, one watt would probably be insufficient. But it would be enough in the oesophagus and the lung. The amount of time required for treatment is also a factor.

Berenbaum: It seems to me that the interest in lasers which produce higher wavelengths is disproportionate. The critical factor in phototherapy is the physical chemistry of the drug and its ability to localize in tumours. Tetra(*m*-hydroxyphenyl)porphyrin is irradiated at 648 nm, which is well below the range we are talking about. With this, we can produce 1 cm thick necrosis in mouse tumours with only 10 J/cm^2 of light. If we were to add a couple of penetration depths to that by increasing the dose of light, we might be able to treat effectively 50% of the human tumours that have not disseminated before the treatment.

Dougherty: There is certainly a trade-off here. I am not sure what the best balance will be. If we find a photosensitizer that works at twice the depth, clinicians will continue to use the laser systems already available.

van den Bergh: Drs Dougherty and Henderson suggested that one could potentiate red light phototherapy using hyperthermia (Henderson et al 1985). At the CHUV Hospital in Lausanne we investigated that with four patients; three with early cancer of the oesophagus and one with early carcinoma of the pharynx. Very preliminary results with three of these patients look promising. We can enhance red light phototherapy with porphyrins (3 mg/kg HpD, 100 mW/cm^2, 20 min) by using a continuous wave YAG laser to produce temperatures of between 41 and 43 °C at the surface of the tumour for the duration of the PDT.

Dougherty: This is interesting because we might be able to use less drug and reduce the problem of skin photosensitivity in addition to obtaining deeper treatment.

Bown: If the temperature is 41 °C at the surface of the tumour, what do you think it is at the deepest point below the surface that you influence biologically?

van den Bergh: That's hard to say, but tissue penetration at 1.06 μm is very good.

Bown: I wonder whether this is a true synergistic effect. The gentle heat may be causing vascular dilatation which makes the tumour more responsive to treatment.

Dougherty: What power density did you use with the YAG laser?

van den Bergh: Up to about 400 mW/cm^2.

Dougherty: That wouldn't go very deep. In our animal experiments we use 600 mW/cm^2 (363 nm) to get 41 °C at the depth of the tumour.

van den Bergh: The penetration also depends on the surface area irradiated. When we used 400 mW/cm^2 in the oesophagus, the temperature at the surface rose to about 43 °C.

Berenbaum: How do you know this is an increased effect? What comparison are you making for these three patients?

van den Bergh: We are comparing red light phototherapy alone with red light phototherapy and additional infrared irradiation. Dr P. Monnier, who treated these patients with PDT and hyperthermia, found that the necrosis started earlier and was deeper than in similar patients treated only with PDT.

Brown: Dr Wilson, does the reflectance spectrometry approach to the noninvasive measurement of drug concentration report just on the surface few mm of the tumour? If so, does that represent a uniform drug concentration throughout the tumour?

Wilson: No. One of the potential advantages of this technique is that it samples light from the same effective volume as the treatment because measurements are made at the photoactivation wavelength. In fluorometry, if a short wavelength is used for excitation, the measurements are limited to the surface, whereas in reflectance spectrometry if the photosensitizer is activated at 630 nm you measure the reflectance spectrum at 630 nm. So the measurement is weighted, in some sense, over the treatment volume. We have not studied the sensitivity of the signal to non-uniform distribution of photosensitizer in the treatment volume.

Brown: How feasible is this technique?

Wilson: It is a relatively straight-forward approach but there are still questions about calibration because the signal strength depends on not only the photosensitizer concentration but also on the optical properties of the tissue.

Brown: Could you calibrate by radiolabelling?

Wilson: No. We intend to calibrate for the tissue optical properties by measuring the total reflectance signal and the reflectance as a function of radial distance along the surface. Thus at one time we shall measure the total reflectance at the photoactivation wavelength, the radial dependence of the local reflectance on the surface at the activation wavelength, and the total reflectance across the photosensitizer absorption band.

References

Bonnett R, Berenbaum M 1989 Porphyrins as photosensitizers. In: Photosensitizing compounds: their chemistry, biology and clinical use. Wiley, Chichester (Ciba Found Symp 146) p 40–59

Henderson BW, Waldon SM, Mang TS, Dougherty TJ 1985 Photodynamic therapy and hyperthermia. In: Jori G, Perria C (eds) Photodynamic therapy of tumours and other diseases. Libreria Progretto, Padua, p 183–193

In vivo transport and pharmacokinetic behaviour of tumour photosensitizers

Giulio Jori

Department of Biology, University of Padua, via Loredan 10, I-35131 Padua, Italy

Abstract. The mechanisms by which photodynamic sensitizers are transported in the bloodstream influence their distribution among normal and tumour tissues, as well as their partitioning among the various compartments of tumour tissues. Column chromatographic analysis and density gradient ultracentrifugation of sera obtained from both patients and experimental animals show that hydrophilic photosensitizers (e.g. haematoporphyrin, and tetrasulphonated porphyrins and phthalocyanines) are largely transported by albumin and globulins and mainly deposited in the vascular stroma of tumours. More hydrophobic photosensitizers (haematoporphyrin oligomers, porphyrin esters, monosulphonated or unsubstituted phthalocyanines) are preferentially incorporated in the lipid core of lipoproteins. Tightly aggregated dyes partly circulate as unbound pseudomicellar structures which can be entrapped in the interstitial regions of the tumour, localize in macrophages, or enter neoplastic cells via pinocytotic processes. Low density lipoproteins (LDL), which are endocytosed by neoplastic cells through a specific receptor-mediated pathway, display the most selective release of photosensitizers to tumours. The binding of the injected photosensitizer to LDL can be enhanced by preincorporation of the dye in liposomal vesicles which are in a quasi-solid state at the body temperature.

1989 Photosensitizing Compounds: their Chemistry, Biology and Clinical Use. Wiley, Chichester (Ciba Foundation Symposium 146) p 78–94

The development of procedures for enhancing the ratio between the photosensitizer concentration in tumours and in peritumoral tissues is an increasingly important area of investigation in the field of photodynamic therapy (PDT). To do this, it is important to exploit certain differences between normal and neoplastic tissues. For example, the presence of specific antigens at the surface of malignant cells can be used by preparing monoclonal antibodies directed against such antigens and covalently coupling them to the photosensitizers (Oseroff et al 1987). It has also been shown that tumour mitochondria are much more efficient than normal mitochondria in accumulating high concentrations of lipophilic cationic dyes, such as rhodamines or kryptocyanines (Powers et al 1986).

A different approach is based on the greater propensity of tumour cells to express receptors for low density lipoproteins (LDL) than most normal cells.

Such receptors specifically recognize the LDL and promote their internalization by cells through the formation of coated pits. LDL are the natural carriers of several hydrophobic photosensitizers in the bloodstream; when LDL are endocytosed by the neoplastic cells, the associated dye is released in the cytoplasm and binds to apolar endocellular matrices (Jori 1987). To enhance the efficiency of LDL as carriers of the photosensitizers, it is necessary to achieve a specific association of the dye with these proteins.

In this paper I shall outline current knowledge of the mechanisms by which photodynamic sensitizers are transported *in vivo* and the relationships between the transport of a photosensitizer and the modes of its localization in tumours. Then I shall describe an experimental approach for enhancing the selectivity of tumour targeting by photosensitizers via the LDL pathway.

Transport of tumour photosensitizers *in vivo*

Protein binding accounts for the transport of a very large proportion of systemically injected porphyrins and their analogues (Jori 1984, Dougherty 1987). At the doses normally used in PDT (3–5 mg/kg body weight), at least 95% of haematoporphyrin (Hp) is complexed by serum proteins. Only highly water-soluble porphyrins, such as uroporphyrin, have a weak affinity for protein binding sites; they circulate almost exclusively in the free state. Figure 1 shows a typical chromatogram obtained by eluting human serum through a column of Sephacryl S-300 (Jori et al 1984). Three protein fractions can be detected, which are mainly constituted by lipoproteins (fast-eluting peak), globulins (intermediate peak) and albumin (slow-eluting peak). The latter accounts for about 50% of the total protein content of the blood. Injected Hp is associated with all three protein peaks, whereas a more polar porphyrin, *meso*-tetra(4-sulphonatophenyl)porphyrin ($TPPS_4$), is preferentially bound by albumin. Haematoporphyrin derivative (HpD) exhibits a chromatographic behaviour very similar to that observed for Hp (Fig. 1). However, a significant fraction of HpD does not bind to proteins; it remains in the blood as a pseudomicellar structure of aggregated material (apparent M_r about 40 000) in which the negatively charged carboxylate functions are clustered at the outer surface and the tetrapyrrolic macrocycle forms the apolar interior.

Thus it appears that the distribution of porphyrins among serum proteins is dependent on their chemical structure. The extension of these studies to a variety of porphyrin derivatives allows one to draw some general conclusions (Kessel 1986a,b, Barel et al 1986). Hydrophilic compounds (Hp and other monomeric components of HpD, tri- and tetrasulphonated derivatives of porphyrins and phthalocyanines) preferentially form non-covalent complexes with albumin and globulins, whereas photosensitizers having a high (>8) octanol/water partition coefficient (oligomeric components of HpD, unsubstituted or mono- and disulphonated porphyrins and phthalocyanines) are largely bound by

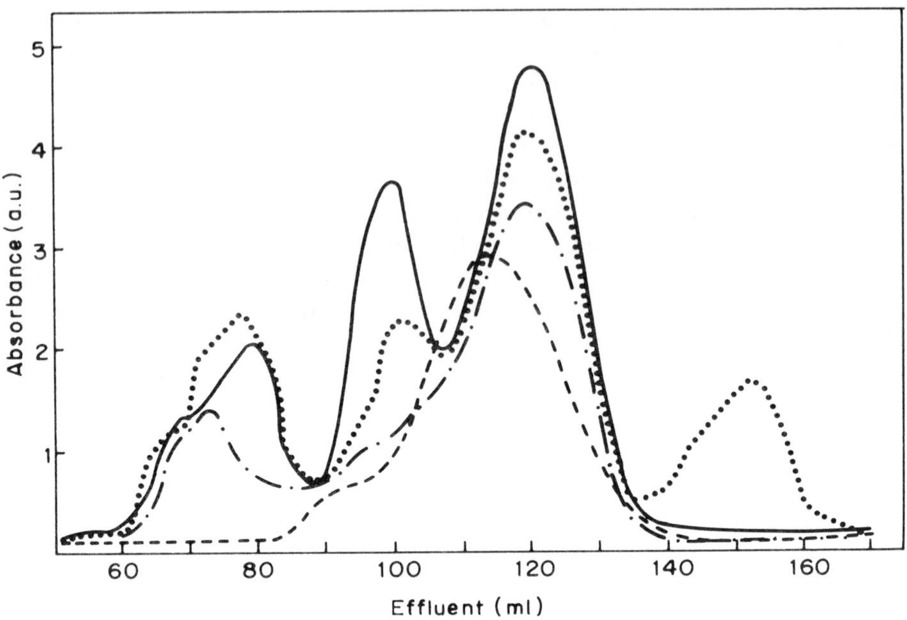

FIG. 1. Elution profiles from Sephacryl S-300 column of photosensitizer-doped human serum. Serum (2 ml) was obtained from patients subjected to PDT 2 h after intravenous injection of photosensitizer (3 mg/kg body weight). The full line represents the protein absorbance at 280 nm. The photosensitizer concentration was monitored by the absorbance at 400 nm for haematoporphyrin (—·—), haematoporphyrin derivative (······) and *meso*-tetra(4-sulphonatophenyl)porphyrin (- - - - -).

lipoproteins. The latter are a heterogeneous group of proteins which can be grossly separated into three main classes by density gradient ultracentrifugation—high density (HDL), low density (LDL) and very low density (VLDL) lipoproteins.

In vitro studies of the interaction of specific porphyrins with isolated proteins (Table 1) indicate that albumin and apo-lipoproteins possess a discrete number of binding sites whose occupancy can be described in terms of a chemical equilibrium between the bound and unbound dye. On the other hand, the binding of photosensitizers with the lipid moiety of lipoproteins is not governed by the law of mass action; it rather reflects a partition of the dye between a lipid and an aqueous phase. Consequently, a relatively large number of photosensitizer molecules can be associated with each lipoprotein.

The similarity of the stability constants for the Hp- or HpD-apoprotein complexes suggests that the distribution of these porphyrins in the serum reflects the relative abundance of the various proteins. The same conclusion holds for the partitioning of hydrophobic photosensitizers among the lipid vesicles of lipoproteins. However, the *in vivo* situation is complicated by the different

TABLE 1 Binding parameters of selected photosensitizers to serum proteins

Protein[a]	*Photosensitizer*[b]	*No. binding sites*	*Dissociation constant*[c] $\times 10^{-6}$ (M^{-1})
Albumin	Hp	1	3.6
	HpD	2	0.2
	TPPS	1	1.9
Apo-HDL	Hp	8	0.7
	Pp	8	6.0
	HpD	10	0.9
Apo-HDL	Hp	12	0.3
	HpD	18	0.7
LDL	Hp	>35	
	Pp	>35	
	HpD	>50	
HDL	Hp	>50	

[a]LDL, HDL, low and high density lipoproteins.
[b]Hp, haematoporphyrin; HpD, haematoporphyrin derivative; Pp, protoporphyrin; TPPS, tetra(4-sulphonatophenyl)porphyrin.
[c]The K_d for holoproteins is not measurable.

clearance rates exhibited by the dyes associated with the different protein carriers. Several reports (Jori 1987, Dougherty 1987) indicate that the fraction of photosensitizers transported by albumin, globulins and apo-lipoproteins is released within about two hours of intravenous injection. LDL also lose the originally bound dye within a few hours, while HDL retain large amounts of the photosensitizer even after 10 days. At the same time, interprotein exchange of the photosensitizer can occur. In particular, we have found that addition of equimolar concentrations of LDL or HDL to an aqueous solution of the albumin–Hp complex leads to an essentially quantitative transfer of the porphyrin to the lipoprotein; the process is very rapid ($t_{1/2} \approx 1$ min) and is probably a consequence of the preferential solubilization of Hp in the lipid core of LDL/HDL, because liposomal vesicles can also efficiently extract the porphyrin from albumin. On the other hand, the incubation of human serum with either LDL–Hp or HDL–Hp systems causes no appreciable delivery of the porphyrin to albumin or globulins. However, the porphyrin becomes gradually redistributed among the various lipoproteins, a steady state being reached when the ratio of Hp bound to HDL, LDL and VLDL coincides with the serum content of the three lipoprotein families.

Influence of the transport mechanism on the mode of photosensitizer localization in tumours

The findings outlined above suggest that photosensitizers circulating in the bloodstream can be operationally divided into three pools (see Table 2):

TABLE 2 Pools of photosensitizer molecules in serum

Pool	*Photosensitizer*	*Carrier*	*Serum half-life*	*Localization in tumours*
Unbound	Aggregated components of HpD	Pseudomicellar structure	2 h	Macrophages, endothelial/neoplastic cells
Weakly bound	Hp, tetrasulphonated porphyrins	Albumin, globulins, apolipoproteins	2 h	Vascular stroma
Strongly bound	HpD oligomers, monosulphonated or unsubstituted porphyrins and phthalocyanines	Lipoproteins (lipid moiety)	7–8 h (LDL) 8 days (HDL)	Neoplastic cells

Abbreviations are defined in Table 1.

(a) photosensitizer molecules which form low affinity bonds with proteins and apoproteins; (b) photosensitizer molecules tightly associated with lipoproteins; and (c) photosensitizer molecules in the free state (e.g. micelle-like structures from HpD oligomers). The pharmacokinetic behaviour of the components of each pool is strikingly different (Jori 1987, West & Moore 1989). Thus the weakly bound and unbound photosensitizers are more readily available for uptake by the various tissues than the strongly bound ones. Because most of the injected drug is weakly bound or free, its rapid disappearance from serum must correspond with the observed maximal accumulation of HpD and other dyes in a variety of normal and tumour tissues 1 to 3 h after administration (Dougherty 1987). About 90% of the intravenously injected photosensitizers disappear from serum with a half-life of about four hours (Nseyo et al 1986). The high concentrations of porphyrin usually recovered from organs such as liver, spleen and kidney strongly support the major role of tissue vascularity in controlling the biodistribution of photosensitizers. In tumour tissues, albumin-carried photosensitizers are mainly released to the interstitial regions, particularly the extracellular matrix (Spikes 1988). At the same time, large porphyrin aggregates can readily interact with macrophages and/or be taken up by endothelial and neoplastic cells of the tumour by direct pinocytosis.

The LDL-bound photosensitizer is delivered to tissues largely by a receptor-mediated endocytotic process. Thus, for these photosensitizers the accumulation is maximal in those tissues which have a large number of LDL receptors (Kessel 1986a). A high LDL receptor activity is typical of several malignant tissues (Spikes & Jori 1987). The bound photosensitizer follows the fate of the LDL; it undergoes internalization by neoplastic cells, and upon lysosomal digestion of the LDL the dye is released into the cytoplasm where it partitions into hydrophobic regions, such as the cytoplasmic and mitochondrial membranes. Consequently, the dye is screened from interaction with the serum proteins which are responsible for its eventual clearance from the organism. The non-exchangeable pool of dye molecules created may be responsible, at least in part, for the prolonged retention of photosensitizers by tumour tissues. The tumour uptake of photosensitizing molecules by the LDL pathway occurs at a relatively low rate and is generally completed in about 12 h.

The dependency of the mode of tumour localization by photosensitizers on the transport mechanism has been investigated by Kessel (1986b) using different sulphonated derivatives of tetraphenylporphyrin. Products with one or two sulphonate groups, which are carried by lipoproteins, are incorporated into neoplastic cells, while their tri- and tetra-sulphonated analogues, which are carried by albumin, are mainly found in stromal elements of the neoplastic tissue. Thus it appears feasible to modulate the endotissular distribution of a photosensitizer in a tumour, and thereby the mechanism of PDT-induced tumour necrosis, by controlling the nature of the dye carriers in the bloodstream.

Finally, the HDL binding probably accounts for the prolonged presence of 5 to 8% of injected photosensitizers in serum (Jori et al 1984, Dougherty 1987) (Table 2). This fraction is cleared from the serum with a half-life of about eight days and thus could represent a persisting pool of material that can localize to tumours by the mechanisms of dye transfer between HDL and LDL and subsequent receptor-mediated endocytosis already described.

The LDL pathway as a tool for increasing the selectivity of tumour targeting by photosensitizers

The importance of the LDL pathway for enhancing the selectivity of photosensitizer delivery to tumours and the efficacy of PDT has been underlined by different investigations:

(a) The administration of a photosensitizer after its *in vitro* complexation with LDL (Jori 1984) leads to a much larger tumour/normal tissues ratio of dye concentration compared with when the same photosensitizer is dissolved in aqueous solution (Table 3). The redistribution of the LDL-associated photosensitizer among other lipoproteins occurs at a slower rate than its accumulation by the neoplastic tissue, which limits the aspecific targeting of normal tissues.

(b) The biodistribution of the hydrophobic constituents of HpD (which are mostly carried by lipoproteins) in experimental animals is correlated with the levels of LDL receptors expressed by the various tissues (Kessel 1986c).

(c) The uptake of albumin- or HDL-bound HpD by human fibroblasts shows a linear increase upon increasing the porphyrin concentration, while the delivery of HpD to fibroblasts via LDL undergoes saturation when the protein receptors at the cell surface are saturated (Candide et al 1986).

(d) The LDL-delivered photosensitizers induce tumour necrosis mainly by direct killing of neoplastic cells, involving both mitochondrial and lysosomal damage (Zhou et al 1988).

TABLE 3 Time-dependency of the ratio between the haematoporphyrin concentration in tumour and selected tissues

Time after injection	*Hp in saline*		*Hp-LDL*	
	Tumour/muscle	*Tumour/skin*	*Tumour/muscle*	*Tumour/skin*
3 h	2.4	1.4	5.6	6.7
24 h	3.4	1.7	9.5	22.4
48 h	3.9	2.2	9.8	20.9
1 week	4.0	2.8	8.4	18.7

Hp was intravenously injected (5 mg/kg body weight) into mice bearing a MS-2 fibrosarcoma transplanted in the right hind leg. Hp recovery from the homogenized tissues was estimated by spectrophotofluorimetric analysis after extraction of the porphyrin with 2% aqueous sodium dodecylsulphate.

Consequently, it appears important to develop procedures of photosensitizer administration which favour its interaction with LDL. Studies from our laboratory (E. Reddi et al, unpublished results) indicate that the selective release of a photosensitizer to lipoproteins can be achieved by its preincorporation into liposomes constituted by phospholipids which are in a quasi-solid state at the body temperature (e.g. distearoyl- or dipalmitoyl-phosphatidylcholine). In particular, the addition of 10–15% cholesterol to the lipid bilayer enhances the efficiency of photosensitizer accumulation by LDL. If the photosensitizer is sufficiently hydrophobic (e.g. phthalocyanines devoid of lateral substituents), there is no detectable leakage of the dye from the LDL lipid core into the bulk aqueous medium. By this procedure, upon intravenous injection of tumour-bearing mice with zinc-phthalocyanine doses as low as 0.14 mg/kg, we obtained dye levels in the tumour in the order of microgrammes per g of tissue as well as an extensive tumour necrosis after PDT (Reddi et al 1987).

Interestingly, the same sequence of events (i.e. highly selective tumour targeting, efficient response of the tumour to PDT, specific killing of neoplastic cells in the tumour tissue) is observed upon injecting a variety of tumour-bearing mice or rats with liposome-bound photosensitizers of different chemical structure, namely Hp, Hp dimethyl ester, Zn-phthalocyanine, Si-naphthalocyanine and tetrapropylporphycene. In all cases, the photosensitizer is selectively transferred to lipoproteins, and a significant fraction (up to 30%) becomes associated with LDL. Thus LDL can act as an endogenous tumour-specific carrier of photosensitizers tailored to any specific phototherapeutic application. For example, the photosensitizer could be chosen on the basis of the optical properties of the tumour and the required penetration depth of the incident light.

Acknowledgements

This work received financial support from the Consiglio Nazionale delle Richerche, Italy, under the special project "Oncologie", contract no: 8701322.44.

References

Barel A, Jori G, Perin A, Romandini P, Pagnan A, Biffanti S 1986 Role of high-, low- and very low-density lipoproteins in the transport and tumour-delivery of hematoporphyrin in vivo. Cancer Lett 32:145–150

Candide C, Morliere P, Maziere JC et al 1986 In vitro interaction of the photoactive anticancer porphyrin derivative Photofrin II with low density lipoproteins and its delivery in cultured human fibroblasts. FEBS (Fed Eur Biochem Soc) Lett 207:133–138

Dougherty TJ 1987 Photosensitizers: therapy and detection of malignant tumours. Photochem Photobiol 45:879–889

Jori G 1984 Pharmacokinetic studies with hematoporphyrin in tumour-bearing mice. In: Jori G, Perria CA (eds) Photodynamic therapy of tumours and other diseases. Libreria Progetto, Padua, p 159–166

Jori G 1987 Photodynamic therapy of solid tumours. Radiat Phys Chem 30:375–380

Jori G, Beltramini M, Reddi E, Pagnan A, Tsanov T 1984 Evidence for a major role of plasma lipoproteins as hematoporphyrin carriers in vivo. Cancer Lett 24: 291–294

Kessel D 1986a Porphyrin–lipoprotein association as a factor in porphyrin localization. Cancer Lett 33:183–188

Kessel D 1986b Photosensitization with derivatives of hematoporphyrin. Int J Radiat Biol 49:901–907

Kessel D 1986c Sites of photosensitization by derivatives of hematoporphyrin. Photochem Photobiol 44:489–494

Nseyo UO, Mang TS, Potter WR 1986 Dihematoporphyrin ether clearance in primate bladder. J Urol 136:1361–1366

Oseroff AR, Ara G, Ohuoha D et al 1987 Strategies for selective cancer photochemotherapy: antibody-targeted and selective carcinoma cell photolysis. Photochem Photobiol 46:83–96

Powers SK, Pribil S, Gillespie GV, Watkins PJ 1986 Laser photochemotherapy of rhodamine-123 sensitized human glioma cells in vitro. J Neurosurg 64:918–923

Reddi E, Lo Castro G, Biolo R, Jori G 1987 Pharmacokinetic studies with Zn(II)-phthalocyanine in tumour-bearing mice. Br J Cancer 56:597–600

Spikes JD 1988 The role of the anatomy, physiology and biochemistry of tumours in the selective retention of sensitizers and the mechanisms of photosensitized tumour destruction. In: Douglas RH et al (eds) Light in biology and medicine. Plenum Press, New York, vol 1:105–113

Spikes JD, Jori G 1987 Photodynamic therapy of tumours and other diseases. Lasers Med Sci 2:3–15

West CML, Moore JV 1989 The photodynamic effect of Photofrin II, hematoporphyrin derivative, hematoporphyrin and tetrasodium-meso-tetra (4-sulfonatophenyl)porphine in vitro: clonogenic cell survival and drug uptake studies. Photochem Photobiol 49:169–174

Zhou C, Milanesi C, Jori G 1988 An ultrastructural comparative evaluation of tumours photosensitized by porphyrins administered in aqueous solution, bound to liposomes or to lipoproteins. Photochem Photobiol 48:487–492

DISCUSSION

Dougherty: Wouldn't you expect any hydrophobic drug to be associated with the hydrophobic, lipophilic areas of the cells and serum, whether or not it was bound to liposomes?

Jori: Yes, but if the drug is hydrophobic when it is administered problems such as precipitation or aggregation may be encountered.

Dougherty: Unlike in the rapidly growing animal tumour models, the fraction of cells that is cycling in human tumours is very small. Are not these cells the only ones with many receptors for low density lipoproteins (LDL)? If so, and if LDL receptors are critical in this transport of the drug, you would not reach most of the tumour cells.

Jori: I don't know. Other studies have been done mainly *in vitro* and show that each type of hyperproliferative cell develops many LDL receptors.

Dougherty: That was my point; the transport depends on proliferation.

Kessel: This doesn't necessarily depend on proliferation; it's a question of membrane turnover, because the LDL is only bringing cholesterol into the cell. Even if the cells are not proliferating rapidly they may still be turning over membranes.

Dougherty: Why are there more LDL receptors in tumour cells than in normal cells?

Kessel: Perhaps because, for unknown reasons, tumours turn over their membranes more frequently than many other normal cells.

Jori: Endothelial cells in tumours have many LDL receptors. Our original aim was to target both endothelial and neoplastic cells and induce vascular and cellular damage at the same time. To our surprise this did not happen. Endothelial cells differ from truly malignant cells in that they perform endocytosis and exocytosis simultaneously.

We are collaborating with Dr W.M. Star. He has been able to observe fluorescence of zinc phthalocyanine (Zn-Pc) in blood vessels.

Dougherty: Was this fluorescence in the supporting stroma?

Jori: That hasn't been determined yet.

Henderson: We tried to put haematoporphyrin (Hp) into liposomes, but after dialysis and high-performance liquid chromatography we found that only the dihaematoporphyrin ether (DHE) contaminant of the Hp solution was in the liposome. All the Hp was outside.

Jori: We find that everything is incorporated in the liposome.

Henderson: Is the Hp in the hydrophilic core?

Jori: It's in the phospholipid bilayer.

Moan: Do you both do the experiment in the same way? Do you prepare the samples by sonication?

Henderson: We followed the regular procedure—sonication followed by exhaustive dialysis.

Dougherty: Hp is water soluble. Why doesn't it come out?

Jori: Hp leaks from the liposomes very slowly, which is why the liposomal preparations of Zn-Pc or Hp dimethyl ester are more stable.

Henderson: But after the freshly prepared liposome preparation was dialysed to remove any free porphyrins, the only component remaining in the liposome was the DHE contaminant.

Jori: That may happen if you do dialysis overnight. After preparing the liposome, we remove any contaminants from the sonicator by centrifugation. The liposomes are then ready for injection.

Henderson: Do you not clean up your preparation to remove unbound material?

Jori: No, there is no free material initially. However, if we incubate the liposome in solution there is a continuous leakage of haematoporphyrin from the phospholipid into the bulk aqueous medium.

Henderson: We observed a preferential leakage of hydrophilic rather than lipophilic compounds.

Jori: That's what we found.

Henderson: Selman et al (1989) incorporated purpurins in liposomes because purpurins also cause strong vascular effects. They did not find any preservation of the vasculature after PDT when purpurins were delivered in this way.

Morgan: In that study we looked at three carriers—liposomes, cremophor emulsion and γ-cyclodextrin—for tin purpurin, which is itself hydrophobic. We have used the cremophor emulsion for four years. If we incubate the sensitizer–emulsion complex with serum *in vitro* the sensitizer binds to the lipoprotein fraction, although we don't yet know the LDL : HDL ratio. If we incubate the sensitizer–γ-cyclodextrin complex with serum *in vitro* the sensitizer binds to the albumin fraction. Using the rat bladder tumour model, we saw a greater *in vivo* tumour response when the delivery agent was the cyclodextrin rather than the emulsion.

meso-Tetraphenylporphyrin tetrasulphonate is delivered by albumin and gets into tumours in high concentrations, but it localizes in the stroma. That might be because it's hydrophilic, and the hydrophobic drug may redistribute in more sensitive areas.

Using electron microscopy we found that delivery by the emulsion results in endothelial cell damage first. Although this study used the emulsion and not liposomes, the dye is transferred to lipoproteins. Therefore there must be some additional factor which we do not understand.

Dougherty: Is cremophor a true emulsion? Is it micellar? Does it scatter light?

Kessel: It is a clear micellar preparation. Fluorescence studies show that the dye is in a hydrophobic region of the micelle.

Morgan: The emulsion, which gives a lower tumour response than the cyclodextrin, gives no skin response at all by the foot pad test. But we do see a foot response with the cyclodextrin, which is presumably delivering more drug everywhere, or perhaps less to the reticuloendothelial system. The results with liposomes are unclear. If we incubate the sensitizer–liposome complex with serum *in vitro* we find the drug in both the albumin and the lipoprotein fractions, and the tumour response is weaker than with either the emulsion or cyclodextrin delivery systems, which doesn't make sense.

Jori: The fact that a drug is in vascular stroma does not necessarily mean that it is a more efficient tumour sensitizer. The overall necrotic effect relies on a complex mixture of factors.

Morgan: We haven't done electron microscopy studies with the cyclodextrin or the liposomes yet. If you increase the concentration of sensitizer do you see the same sequence of events?

Jori: We used two concentrations of Zn-Pc, 0.25 and 0.5 mg/kg, and saw the same sequence of events. Zn-Pc produces the same pattern of events in two different tumour models. There is no initial change in the blood capillaries, and the dilatation followed by blockage of blood vasculature that we observed initially with HpD is not seen. The liposomes are delicate; their properties are critical and the preparation procedure might be important.

Dougherty: How much zinc phthalocyanine is incorporated into the liposomes?

Jori: We can achieve a molar ratio of about one phthalocyanine molecule per 30 molecules of phospholipid.

Brown: What concentration does that give in solution?

Jori: Less than 1 mg/ml.

Morgan: Do you use pyridine to solubilize the phthalocyanine?

Jori: Yes. The pyridine is removed by dialysis. Zn-Pc is fairly stable in the liposomes and doesn't leak into the medium.

Morgan: Using THF we get higher concentrations.

Ullrich: Josh Fidler has activated macrophages with liposomes and muramyl dipeptide and the macrophages phagocytose the liposomes (Fidler & Schroit 1984). Are you possibly doing the opposite—directing a drug to a macrophage, hitting that macrophage with light and inactivating macrophages inside the tumour? Might that explain why you are not killing the tumour cells? If the size and composition of the liposome are altered the amount of phagocytosis changes. This may explain the differences.

Jori: The liposome we use has a half-life in serum of a few minutes; it fuses with lipoproteins. We tried many different preparations and found that the radius of curvature of the liposome is critical. For an easy interaction with lipoproteins, the fluidity of the phospholipid bilayer is also important. When we prepared unilamellar liposomes with dimyristoyl phosphatidylcholine rather than dipalmitoyl phosphatidylcholine, which has one carbon atom more in the hydrocarbon chain, we found that the Zn-Pc was associated with all types of serum protein; there was much less selectivity. This occurs with all phospholipids that are in a fluid state at 37 °C. Therefore we must use phospholipids that are in a solid state at this temperature. Most studies with liposomes have been done with lecithin phospholipids, which are a mixture of different phospholipids and are fluid at 37 °C.

Morgan: Have you tried preblocking the reticuloendothelial system with blank liposomes so that the macrophages are saturated and don't take up the drug?

Jori: This might be done. But it is clear that we are loading only the malignant cells.

Kessel: Table 1 shows the results of separating the different lipoprotein species of human serum in an ultracentrifuge. This is a static determination; we don't have the option of seeing what happens with time. The protein fraction consists mainly of albumin. LDL and HDL contain similar amounts of HpD (the tumour-localizing fraction). Tetra(*m*-hydroxyphenyl)porphyrin (*m*-THPP) binds more to LDL than to HDL, so this dye might localize well in tumours. *N*-aspartyl chlorin e_6 (NPe_6) has four carboxyl groups on one side and is mainly bound to protein. This agent tends to associate with lysosomes whereas HpD binds to mitochondria. The tetraphenylporphyrin sulphonates bind more to albumin as the degree of sulphonation increases, and the tetrasulphonate localizes in stroma. Thus one can direct the dyes where they are wanted. If one were to put some of these drugs that bind albumin into liposomes and direct them at LDL, a completely different pattern of binding might result.

Moan: Have you checked the redistribution at various times? The pattern of distribution of Photofrin II in a patient's serum appears to be the same a few hours and a few days after injection.

Kessel: This is all *in vitro* work.

Brown: We took LDL- or HDL-bound Photofrin II and added it back to whole serum in amounts that didn't disturb grossly the amount of lipoprotein present. Redistribution was very fast.

Moan: Did you check the distribution among the serum proteins at different times after injection?

Brown: That's a different experiment. We have looked at patients at various times from immediately after injection to several weeks later. Although the absolute levels of drug drop dramatically, the proportions in HDL and LDL stay almost constant. But that doesn't necessarily prove exchange.

Bonnett: Is there aggregation of the drug?

Moan: In some fractions, notably the albumin fraction, I think there is heavy aggregation, even at low concentrations, whereas in the LDL fraction there is no aggregation.

TABLE 1 (Kessel) Steady-state binding of dyes to plasma fractions

	Percent dye binding to each fraction			
Dye	*VLDL*	*LDL*	*HDL*	*Protein*
Hp	1.5	14	82	2.5
HpD(tlf)	3	44	58	2
Pp	5	19	71	5
P-C_6-P	0	5	93	2
m-THPP	4	80	16	0
e_6	0	1	58	41
NPe_6	0	2	29	70
$TPPS_1$	15	28	57	0
$TPPS_2$	0	20	65	15
$TPPS_4$	0	3	14	87

Hp, haematoporphyrin; HpD(tlf), tumour-localizing fraction of HpD; Pp, protoporphyrin; P-C_6-P, diporphyrin with six methylenes between the porphyrin rings; *m*-THPP, tetra(*m*-hydroxyphenyl)porphyrin; e_6, chlorin e_6; NPe_6, *N*-aspartyl chlorin e_6; TPPS, tetraphenylporphyrin sulphonate (subscript denotes number of SO_3H groups). Human plasma was incubated for 30 min at 37 °C with 5 μM dye, and the fractions were then separated by density-gradient centrifugation.

Kessel: That is probably true.

Dougherty: Dr Jori showed an unbound fraction. Do you see an unbound fraction in your *in vitro* experiments, Dr Kessel?

Kessel: We do not add enough dye to see a fraction that is not bound to some macromolecule.

Jori: We detected the unbound fraction with an injection of only 2.5 mg/kg of sensitizer.

Dougherty: Does this unbound fraction disappear quickly?

Jori: Yes, the serum half-life is 2–3 h.

Moan: Is there a redistribution among different cell proteins?

Jori: The redistribution is mainly among lipoproteins. The distribution pattern among lipoproteins becomes constant after some time.

Dougherty: But after a long time, at least with Photofrin II, the drug is mainly bound to HDL. That could represent enterohepatic circulation, for example. Might that account for some of the long-term cutaneous photosensitivity? This constant low level of circulating porphyrin is similar to what occurs in porphyrias. Protoporphyrin, for example, also binds preferentially to LDL and must be circulating in that way.

Jori: Reyftmann et al (1984) found that porphyric patients have a large amount of protoporphyrin associated with LDL.

Berenbaum: The fact that you did not see endothelial damage is difficult to understand. You pointed out that there are LDL receptors on the endothelium. Your suggestion to account for the lack of damage is that endocytosis and exocytosis occur simultaneously. In our experiments with HpD in normal brain in mice (Berenbaum et al 1986), the first evidence of damage, within two hours, was on the vascular endothelium. There is a blood–brain barrier, constituted by the endothelial cells and the tight junctions that seal them off, and no apparent mechanism for HpD transport. So the damage is in the endothelial cells. The endothelium remains sensitive even three months after the injection; the LDL receptor–drug complex is extremely stable. I would ascribe prolonged brain and skin sensitivity not to the continued circulation of the drug but to the presence of receptors, possibly in cell membranes, which are not turning over.

Dougherty: The long retention is not related to the receptor; that's turning over rapidly. The dye is endocytosed and binds to some intracellular structure.

Jori: We have no proof for the exocytosis hypothesis. The lack of damage may be also related to the finding by Charles Gomer (unpublished paper, International Conference on Photodynamic Therapy, London, July 1988) that endothelial cells have a unique pattern of stress proteins after photosensitization which differs from that of fibroblasts or malignant cells.

Bown: What is known about the distribution of liposome-delivered drugs between tumours and normal tissue?

Jori: We use a fibrosarcoma which is grown intramuscularly in the right hind leg of BALB/c mice. The left hind leg can be used as a control.

The liposome-delivered zinc phthalocyanine is distributed between the tumour and the left healthy leg in the ratio 7 to 9 : 1 after 24 h.

Bown: What ratio would you get without using liposomes?

Jori: With zinc phthalocyanine that is impossible. With haematoporphyrin, which can be injected both in liposomes and in aqueous solution, the tumour to muscle ratio is approximately doubled when the drug is delivered in liposomes (Table 3). But with liposome-delivered zinc phthalocyanine we do much better. We think this is because haematoporphyrin leaks from the liposome and the LDL whereas the Zn-Pc complexes are stable.

Bown: Is the variation in distribution matched by a difference in necrosis?

Jori: Yes; with liposome-delivered Zn-Pc, when we irradiate the normal leg with the dose rates and total light doses necessary for tumour necrosis (180 mW/cm^2 and 300 J/cm^2), we do not observe significant photosensitization.

Bown: If you were doing that with dye in aqueous solution you would expect to see damage in the muscle under similar conditions.

Jori: Yes, we do.

Brown: Photofrin II is mostly bound to HDL. If you culture cells in the absence of lipoproteins, you get a far higher uptake of drug than if lipoproteins are present. To what extent can HDL deliver drug to the membrane of the cell and that then be taken up non-specifically?

Jori: Tumour cells don't have receptors for HDL. There is some non-specific fluid endocytosis, but we have not determined the contributions of the various pathways to the localization of phthalocyanines in the tumour. This has been done to some extent by Candide et al (1986) with HpD. By adding increasing amounts of HpD bound to LDL they observed saturation of the receptors. On the other hand, if you add HpD bound to albumin or HDL there is a linear increase in the uptake which depends on the concentration of the photosensitizer. This shows that both pathways may contribute. But one should titrate the number of the receptors to estimate the true contribution of LDL.

Dougherty: They did that experiment.

Jori: In vitro it's possible, but *in vivo* it's very difficult. The LDL pathway is not the only route by which photosensitizers are transported and delivered to tissues (Morliere et al 1987). Phthalocyanine is bound to HDL in large amounts, because liposomes deliver the drug to all lipoproteins, of which HDL is the most abundant. But by restricting the binding to lipoproteins one obtains a better localization and tumour response; this is what I was trying to demonstrate. If we could find a delivery system which releases all the drug to LDL, we might achieve maximum selectivity, but it doesn't work.

Moan: Why not remove serum from the patient, isolate the LDL by centrifugation and use them?

Dougherty: That's difficult and takes a long time.

Brown: And the drug redistributes when you replace the LDL.

Kessel: What proportion of the LDL would you need to isolate?

Jori: About eighty molecules of Hp bind to each lipoprotein molecule.

Dougherty: You would not need all the LDL.

Jori: If you reinject LDL complexed with a photosensitizer you have competition with endogenous LDL whose biosynthesis is stimulated by the removal. We are trying to prepare some synthetic LDL, obtain the apoprotein and then reassociate it with liposomes. We hope to minimize the transfer of the drug from the synthetic LDL to other lipoproteins.

Brown: Could you use covalent binding of the drug to the lipoprotein?

Jori: That may interfere with the lipoprotein–receptor interaction.

Moan: The use of reconstituted LDL is also important for other drugs; cytostatic drugs stick better to reconstituted lipoproteins.

Jori: It's not difficult to delipidate the lipoproteins, but it's very difficult to obtain a stable reconstitution. If we succeed, it might be sensible to prepare the complex *in vitro* and inject it.

Dougherty: Redistribution might still be a problem.

Moan: Drugs bound to reconstituted LDL do not redistribute (Masquelier et al 1986, Lindberg 1987).

Kessel: For efficient binding to LDL the drug must be relatively hydrophobic. Iwanik et al (1984) attempted to circumvent a mode of drug resistance in mouse tumour cells by binding an analogue of adriamycin to LDL. The binding was so inefficient that it was only feasible to do the appropriate study in cell culture; the volume of solution necessary for injection into a mouse would weigh more than the mouse. Fortunately, porphyrins can be quite hydrophobic.

Dougherty: Recent studies used liposome-delivered adriamycin. How does that work?

Henderson: Adriamycin was incorporated in the hydrophilic core of the liposomes.

Jori: That experiment used large vesicles with relatively large endoliposomal aqueous cores. Can it be done with other vesicles?

Henderson: It depends on the liposome preparation, but it is possible.

Kessel: But unfortunately the drug still goes to the liver.

Berenbaum: Regional perfusion might be useful. One can virtually isolate some organs, especially limbs, and circulate the complexes through the region containing the tumour with very little exchange with other regions.

References

Berenbaum MC, Hall GW, Hoyes AD 1986 Cerebral photosensitization by haematoporphyrin derivative. Evidence for an endothelial site of action. Br J Cancer 53:81–89

Candide C, Morliere P, Maziere JC et al 1986 In vitro interaction of the photoactive anticancer porphyrin derivative Photofrin II with low density lipoproteins and its delivery in cultured human fibroblasts. FEBS (Fed Eur Biochem Soc) Lett 207:133–138

Fidler IJ, Schroit AJ 1984 Synergism between lymphokines and muramyl dipeptide encapsulated in liposomes: in situ activation of macrophages and therapy of spontaneous cancer metastases. J Immunol 133:515–518

Iwanik MJ, Shaw KV, Ledwith BJ, Yanovich S, Shaw JM 1984 Preparation and interaction of a low-density lipoprotein:daunomycin complex with P388 leukemic cells. Cancer Res 44:1206–1215

Lindberg B 1987 Preparation of drug–low density lipoprotein complexes for delivery of antitumoral drugs via the low density lipoprotein pathway. Cancer Res 47:4105–4108

Masquelier M, Vitoli S, Peterson C 1986 Low-density lipoprotein as a carrier of antitumoral drugs: in vivo fate of drug–human low density lipoprotein complexes in mice. Cancer Res 46:3842–3847

Morliere P, Kohen E, Reyftmann JP et al 1987 Photosensitization by porphyrins delivered to L cell fibroblasts by human serum low-density lipoproteins. A microspectrofluorometric study. Photochem Photobiol 46:183–191

Reyftmann JP, Morliere P, Goldstein S, Santus R, Dubertret L, Lagrange D 1984 Interaction of human serum low density lipoproteins with porphyrins: a spectroscopic and photochemical study. Photochem Photobiol 40:721–729

Selman SH, James HR Jr, Keck RW, Garbo GM, Morgan AR 1989 Blood flow changes with purpurin-PDT: does the vehicle make a difference? Proc International Society for Optical Engineering (SPIE) 1065, in press

Intracellular localization of photosensitizers

J. Moan*, K. Berg*, E. Kvam*, A. Western*, Z. Malik‡, A. Rück°, and H. Schneckenburger°

Institute for Cancer Research, Montebello, 0310 Oslo 3, Norway and ‡*Bar Ilan University, 53100 Ramat Gan, Israel* and °*Universität Ulm, Postfach 4066, D-7900 Ulm, Federal Republic of Germany*

Abstract. The intracellular localization of photosensitizers can be studied by different methods. One method involves homogenization of the cells followed by differential ultracentrifugation which leads to fractions enriched in nuclear, mitochondrial, and microsomal material as well as a supernatant fraction. More detailed information can be obtained by electron microscopy of cells exposed to light in the presence of photosensitizers. This method is based on the assumption that damage is primarily induced at intracellular sites where the concentration of photosensitizer is high. By irradiating the cells at 6 °C, where biochemical reactions are slow, and then incubating them for different times at 37 °C, it is possible to follow the development of damage. The amount of photosensitized damage to enzymes or cell functions whose localization in the cells is known gives information about the intracellular localization of the sensitizer. Fluorescence microscopy is the most direct method and is widely applicable because most photosensitizers fluoresce. Lipophilic dyes generally localize in membrane structures. In future more attention should be paid to the localization of dyes in lysosomes, as suggested by early reports. Mitochondria, the endoplasmic reticulum and nuclear membrane are other important loci for intracellular localization of sensitizers.

1989 Photosensitizing Compounds: their Chemistry, Biology and Clinical Use. Wiley, Chichester (Ciba Foundation Symposium 146) p 95–111

Knowledge about the intracellular localization of photosensitizing drugs is important for several reasons: (1) The localization is related to the mechanism by which the sensitizers are taken up or produced by the cells. For therapeutic applications it may be of interest to modify this uptake or production; (2) Most biological photosensitizers act via the singlet oxygen ($^1O_2$) pathway. The range of action of this species in cells is short (about 0.1 μm). Thus, cellular structures close to high sensitizer concentrations will be preferentially harmed by light. Consequently, the intracellular localization of a sensitizer may be closely related to its mechanism of action. It may be of interest to avoid damage to certain structures; for example, during photodynamic therapy (PDT) one does not want to induce mutations in surviving cells which might lead to carcinogenesis. Therefore one should avoid sensitizers that localize in the nucleus. If one wishes to hit a particular cell function, such as DNA synthesis in psoriasis, one should look for a drug that localizes where that function occurs.

Photosensitizers may also be useful in the study of basic intracellular functions. Some sensitizers localize selectively in mitochondria or lysosomes. They may be used to induce damage to these organelles. In this paper we shall illustrate how sensitizers can be used to study the organization of DNA in the nucleus. Four approaches can be used to investigate the intracellular localization of photosensitizers: subcellular fractionation; studies of sensitizing effects on site-specific enzymes or cell functions; morphological studies of cells exposed to light in the presence of the sensitizers; and fluorescence microscopy.

It should be noted that cell-bound sensitizers are much more efficient in cell inactivation than unbound extracellular sensitizers (Moan et al 1982a). Furthermore, the efficiency of a cell-bound sensitizer is strongly dependent on its localization within a cell: hydrophilic sensitizers are far less efficient than lipophilic ones with a comparable singlet oxygen yield (Moan et al 1987). During exposure to light the intracellular localization may alter, giving rise to a change in the quantum yield of cell inactivation (Moan 1988).

Subcellular fractionation

Subcellular fractionation involves homogenization of the cells or tissue and isolation of fractions by means of differential centrifugation. In this way, fractions enriched in nuclear, mitochondrial, lysosomal and microsomal (endoplasmic reticulum) material, as well as a soluble fraction, can be obtained. The fractions are not completely clean, and enzyme assays should always be used to check the presence of known markers for each fraction. Notably, the nuclear fraction may contain a significant amount of membrane material (Cozzani et al 1981). Another limitation of the method is that during the homogenization some of the organelles may release part of the sensitizer bound to them. This may explain the large amount of sensitizer found in the soluble fraction in some experiments (see below).

Data on the intracellular localization of different sensitizers obtained by this procedure are shown in Table 1. In the work of Sandberg & Romslo (1981, 1982) accumulation of porphyrins in rodent livers was induced by feeding the animals with toxic compounds (griseofulvin and hexachlorobenzene) that disturbed the endogeneous haem synthesis. With some selectivity, the lipophilic dye protoporphyrin accumulated in mitochondria whereas the hydrophilic dye uroporphyrin was found to accumulate in lysosomes. The general trend for haematoporphyrin, which is intermediate between protoporphyrin and uroporphyrin with regard to lipophilicity, was that a larger fraction of it was found associated with mitochondria than with lysosomes. It cannot be excluded

TABLE 1 Subcellular fractionation of sensitizers

			Percentage of dye in each fraction					
Sensitizer[a]	*Time*[b]	*Cells*[c]	*Nuclear*	*Mitochondrial*	*Lysosomal*	*Microsomal*	*Soluble*	*Reference*
Hp	1 h (ip)	Rat hepatocytes	8	32	23	21	16	Cozzani et al (1981)
Hp	6 h (ip)	Rat hepatocytes	19	28	12	22	19	Cozzani et al (1981)
Hp	3 h (iv)	Ascites	84	5	2	1	8	Cozzani et al (1981)
Hp	3 h (ip)	Ascites	6.5	1.5	0.5	2.0	89.5	Cozzani et al (1981)
Hp	24 h (ip)	Rat tumour	15[d]	2[d]		4[d]	79[d]	Winkelman (1961)
Pp	Gf	Mouse livers	27	33	10	6	24	Sandberg & Romslo (1981)
Uro	Hex	Rat livers	34.5	17.5	33.5	3.5	11.0	Sandberg & Romslo (1982)
HpD		V79	4	14	n.d.	19	63	Gomer (1978)

[a]Pp, protoporphyrin; Uro, uroporphyrin; HpD, haematoporphyrin derivative.
[b]Time between injection (i.p. or i.v.) of sensitizer and fractionation. Gf, griseofulvin-fed mice; Hex, hexachlorobenzene-fed rats.
[c]Ascites, Yoshida hepatoma AH-130 in rat; Rat tumour, Walker carcinosarcoma 256 homogenates.
[d]Data for the fluorescence assay only, with the background level subtracted. n.d., not determined.

that the haematoporphyrin preparations used in the cited investigations contained trace amounts of lipophilic porphyrins and that the cellular uptake of these impurities dominated that of haematoporphyrin (Moan et al 1982a).

The relatively large amount of sensitizers found in the nuclear fraction is probably related to membranes present in this fraction because in fluorescence microscopy studies none of the sensitizers shows any detectable localization inside the nucleus (see below).

It might seem from the data in Table 1 that the route of administration is of great importance for the intracellular localization of haematoporphyrin in ascites tumour cells. However, the extracellular concentration of the dye was more than 1000 times higher in the case of i.p. administration than in the case of i.v. administration. Therefore, saturation phenomena may explain the differences. The time between injection of the drug and the analysis seems to be of some importance. The amount of porphyrin in the lysosomal fraction decreases with increasing time (Cozzani et al 1981).

Studies of sensitizing effects

Such studies may be carried out in two ways:

(a) It is possible to expose the cells or tissues to the photosensitizers, fractionate the material, and study the effects of light on the isolated organelles or enzymes. Such studies have shown that lipid-rich microsomal membranes, and cytochrome *P*-450 embedded therein, from livers of rats injected with haematoporphyrin derivative (HpD) are damaged by light (Das et al 1985). Hilf et al (1984) showed that the cytosol enzyme pyruvate kinase (PK) and, notably, mitochondrial enzymes were damaged when mitochondria and cytosol from RB230AC mammary adenocarcinomas of rats injected with HpD were exposed to light. Thus HpD seems to be bound to the endoplasmic reticulum membranes as well as to mitochondrial membranes. The effect on mitochondrial enzymes in the above mentioned tumours increased with the time between HpD injection and analysis (up to about 70 h). Furthermore, HpD seemed to be present in the cytosol of the tumour cells for several hours after the injection;

(b) It is possible to study the effect of photodynamic treatment on site-specific functions in intact cells. DNA synthesis is one such site-specific function which takes place at the nuclear membrane, at least in some cell lines including the NHIK 3025 line (data not shown). The fact that DNA synthesis in some cells is extremely sensitive to photodynamic treatment with HpD (Moan et al 1983, Kessel 1986) indicates that: (i) At least one fraction of HpD accumulates at the nuclear membrane. This is in agreement with fluorescence microscopic evidence (Moan et al 1982b); and (ii) The DNA synthesis or factors upon which the DNA synthesis is dependent are located close to the nuclear membrane. Increasing the incubation time leads to an increase in the size of the intracellular pool of non-exchangeable HpD, but no significant increase in the exchangeable pool. Furthermore, membrane damage is more pronounced in cells exposed to light after a short incubation with HpD without washing than in washed cells or cells after a long (≈ 24 h) incubation. An interpretation of this is that the exchangeable pool is located in the outer cell membrane, where it rapidly accumulates, while the non-exchangeable pool is located inside the cells, from where it cannot be easily removed by washing with media containing serum and where it sensitizes damage to cell functions such as DNA synthesis and respiration. For HpD the exchangeable pool contains mainly monomers—Hp, haematoporphyrin vinyl deuteroporphyrin (HVD) and protoporphyrin (Pp)—whereas the non-exchangeable pool contains dimers, polymers and aggregates.

On the basis of fluorescence spectroscopy it was concluded that mesoporphyrin IX, a relatively hydrophobic dye, accumulated rapidly at cellular loci with a high dielectric constant (≈ 20) and more slowly at loci with a lower dielectric constant (≈ 10) (Kessel & Kohn 1980). However, one should note that binding to a water-soluble molecule such as albumin may result in shifts of the fluorescence spectra, similar to the shifts seen upon binding to lipid structures.

Irradiation of sensitizer-loaded cells with microbeams indicates that the plasma membrane and the mitochondria are sensitive targets (Moreno & Salet 1985).

A crucial question is whether a sensitizer localizes in the nucleus or not. DNA itself may be used as a probe for this. A sensitive method is to measure the number of alkali-labile sites introduced into DNA. Figure 1 shows that by X-rays or PDT with the hydrophilic sensitizer tetraphenylporphyrin tetrasulphonate ($TPPS_4$) it is possible to damage all the DNA in the cells. On the other hand, PDT with the hydrophobic dyes Photofrin II (PII) and tetra(3-hydroxylphenyl)porphyrin (3-THPP) introduces damage in only a fraction of the DNA, even after very high exposures. The damaged fraction is supposedly that located close to the nuclear membrane. Thus these sensitizers seem to be completely absent from the nucleus. According to recent models of DNA structure in interphase cells, loops of DNA are attached to the nuclear scaffold. Our photosensitization experiments indicate that these loops have an average length of about 160 kilobases.

Enzymes were liberated from the lysosomes of mouse L-cells upon treatment with methylene blue or haematoporphyrin and light (Santus et al 1983). This indicates a localization of sensitizer in lysosomes or lysosomal membranes.

Morphological studies

Morphological studies of cells exposed to PDT give information about the intracellular localization of the sensitizers, provided the time between light exposure and fixation of the cells is so short that general cell deterioration has not had time to take place. Thus, Fig. 2 shows that PDT with Photofrin II damages the spindle apparatus in mitotic cells.

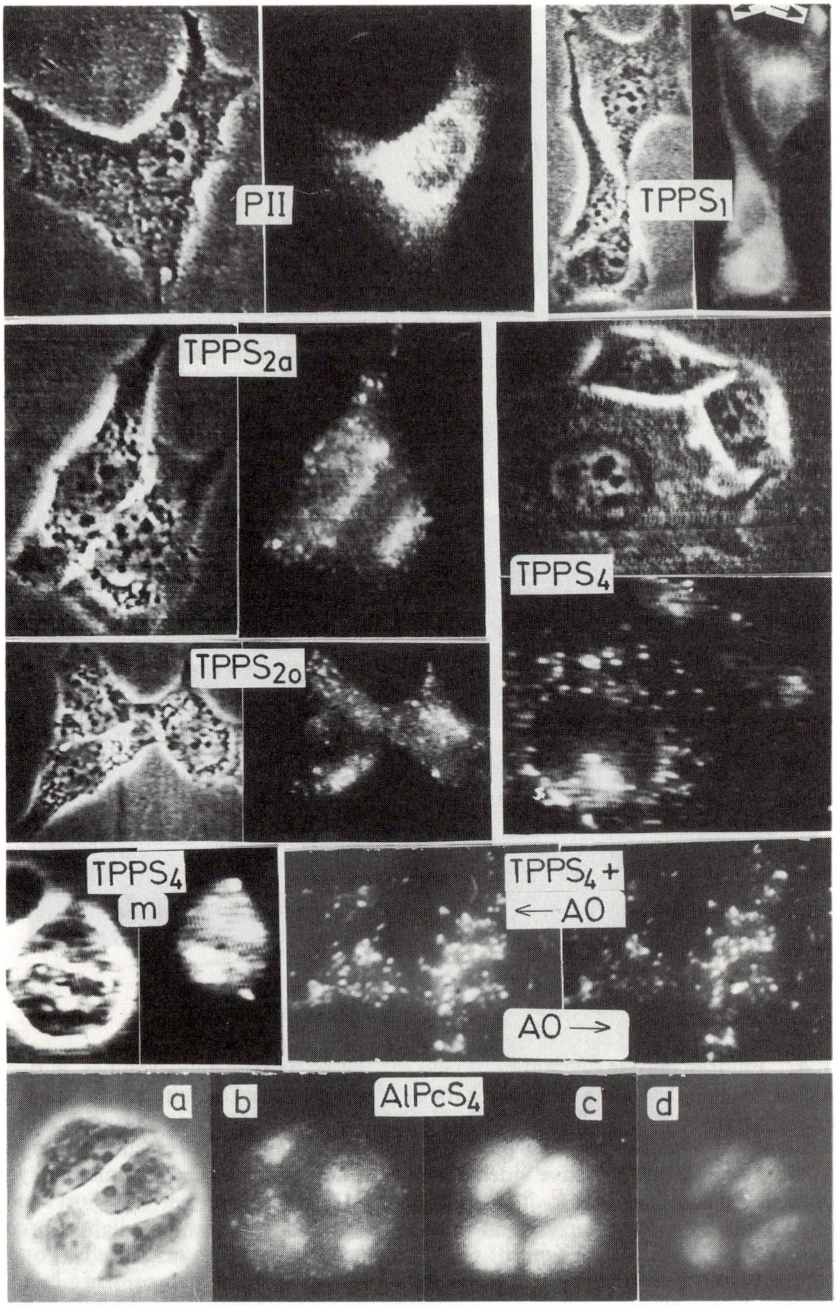
PII
TPPS1
TPPS2a
TPPS4
TPPS2o
TPPS4
m
TPPS4 +
← AO
AO →
a
b
AlPcS4
c
d

Fluorescence microscopy studies

This method gives the most direct picture of the intracellular localization of fluorescing sensitizers. General findings, as illustrated in Figs 4 and 5, are that lipophilic porphyrin and phthalocyanine sensitizers bind to plasma, nuclear and mitochondrial membranes but do not localize in the cell nucleus (Berns et al 1982, Moan et al 1982b, Paquette et al 1988). On the other hand, merocyanine 540, which is also lipophilic, has been reported not to penetrate intact cells but to bind selectively to the plasma membrane (Schlegel et al 1980).

In an attempt to evaluate the influence of lipophilicity on the intracellular localization of structurally related sensitizers we studied a series of tetraphenylporphyrin sulphonates, with one ($TPPS_1$), two adjacently located ($TPPS_{2a}$), two oppositely located ($TPPS_{2o}$), and four ($TPPS_4$) sulphonate groups (Fig. 5). The lipophilicity of these dyes decreases in the mentioned order. $TPPS_1$ was found to localize in the plasma membrane and in the perinuclear region. However, upon removing the membrane-bound dye by washing with a serum-containing medium, distinct grains of fluorescence were seen in the cytoplasm. Such grains appeared even more distinctly in cells incubated with $TPPS_{2o}$, $TPPS_{2a}$ and $TPPS_4$ (Fig. 5). These grains probably correspond to lysosomes, as indicated by the micrographs of cells stained with both $TPPS_4$ and acridine orange (AO). AO is known to fluoresce red when accumulated in lysosomes (Septimus et al 1983). When stained with AO the NHIK 3025 cells showed yellow fluorescence in the nucleus and red spots of fluorescence in the cytoplasm. By using a filter system for excitation and fluorescence emission in the fluorescence microscope, we were able to block the nuclear fluorescence and the $TPPS_4$ fluorescence completely and photograph the red AO fluorescence from the lysosomes alone (Fig. 5). Another filter system also allowed fluorescence from $TPPS_4$ to reach the video camera. Exactly the same pattern, grain by grain, was seen in the two cases. Thus, AO and $TPPS_4$ are localized in the same grains—presumably the lysosomes. An indication of endocytotic uptake of these dyes is shown by the arrows in the micrograph for $TPPS_1$ which point to brightly fluorescent pseudopods or ruffling membranes of the cells, regions which are known to be involved in endocytosis and from which endocytotic channels lead into the cytoplasm.

FIG. 5. Phase contrast and fluorescence micrographs of NHIK 3025 cells incubated with the sensitizers indicated on the figure: Photofrin II (PII; 6 μg/ml), $TPPS_1$ (5 μg/ml), $TPPS_{2a}$ (3 μg/ml), $TPPS_{2o}$ (30 μg/ml), $TPPS_4$ (75 μg/ml), aluminium phthalocyanine tetrasulphonate ($AlPcS_4$) (75 μg/ml). Incubation was 40 h for $TPPS_{2o}$ and 18 h for the other sensitizers. The medium was E2a with 3% serum. The micrographs were recorded by means of a red-sensitive video camera. $TPPS_4$, m shows a mitotic cell. $AlPcS_4$, a is a phase contrast picture, while b, c and d correspond to fluorescence micrographs taken after 0 s, 30 s and about 200 s exposure to the 366 nm exposure light of the microscope.

In mitotic cells the nuclear membrane is absent. However, sensitizers do not seem to localize homogeneously in such cells. Thus the fluorescence from $TPPS_4$ is significantly lower in the equatorial plane, where the chromosomes assemble, than in the rest of the mitotic cell (Fig. 5, $TPPS_4$, m).

The fluorescence of moderate or high concentrations of phthalocyanines in cells increases strongly during light exposure (Figs 4 and 5). Concomitantly, the grainy pattern gradually disappears, indicating lysosomal rupture. The fluorescence increase is not seen in cells with low dye concentrations and is probably due to disaggregation of the dye. Furthermore, as seen from the micrographs of cells containing aluminium phthalocyanine tetrasulphonate ($AlPcS_4$), a relocalization seems to occur in certain cases: during irradiation, fluorescence appears in the nuclear region. Cells loaded wih $TPPS_4$ and exposed to a light dose that inactivated 50% of them and then incubated at 37 °C for 1 h showed a striking change in the fluorescence pattern: the grains had disappeared (lysosomal rupture) and fluorescence was clearly seen in the nucleus. This is in agreement with the observation that all the DNA in the cells can be damaged by light exposure in the presence of this dye (Fig. 1).

Acknowledgements

This work was supported by Hydro Pharma AS, and by The Norwegian Cancer Society. The authors appreciate critical reading of the manuscript by Professor Claude Rimington, F.R.S.

References

Berg K, Moan J 1988 Photodynamic effects of Photofrin II on cell division in human NHIK 3025 cells. Int J Radiat Biol 53:797–811

Berns MV, Dahlman A, Johnsen FM et al 1982 In vitro cellular effects of haematoporphyrin derivative. Cancer Res 42:2325–2329

Cozzani I, Jori G, Reddi E et al 1981 Distribution of endogenous and injected porphyrins at the subcellular level in rat hepatocytes and in ascites hepatoma. Chem Biol Interactions 37:67–75

Das M, Dixit R, Mukhtar H, Bickers DR 1985 Role of active oxygen species in the photodestruction of microsomal cytochrome P-450 and associated monooxygenases by hematoporphyrin derivative in rats. Cancer Res 45:608–615

Evensen JF, Malik Z, Moan J 1988 Ultrastructural changes in the nuclei of human carcinoma cells after photodynamic treatment with hematoporphyrin derivative and with tetrasodium-meso-tetra-(4-sulphonatophenyl) porphine. Lasers Med Sci 3:195–206

Gomer CJ 1978 Evaluation of in vivo tissue localization properties and in vitro photosensitization reactions of hematoporphyrin derivative. PhD Dissertation, State University of New York at Buffalo.

Hilf R, Smail DB, Murant RS, Leakey PB, Gibson SL 1984 Hematoporphyrin derivative-induced photosensitivity of mitochondrial succinate dehydrogenase and selected cytosolic enzymes of RB230AC mammary adenocarcinomas of rats. Cancer Res 44:1483–1488

Kessel D 1986 Sites of photosensitization by derivatives of hematoporphyrin. Photochem Photobiol 44:489–493

Kessel D, Kohn KI 1980 Transport and binding of mesoporphyrin IX by leukemia L 1210 cells. Cancer Res 40:303–307

Moan J 1988 A change in the quantum yield of photoinactivation of cells observed during photodynamic treatment. Lasers Med Sci 26:93–97

Moan J, Christensen T, Sommer S 1982a The main photosensitizing components of hematoporphyrin derivative. Cancer Lett 15:161–166

Moan J, Johannessen JV, Christensen T, Espevik T, McGhie JB 1982b Porphyrin-sensitized photoinactivation of human cells in vitro. Am J Pathol 109:184–192

Moan J, McGhie J, Jacobsen PB 1983 Photodynamic effects on cells in vitro exposed to hematoporphyrin derivative and light. Photochem Photobiol 37:599–604

Moan J, Peng Q, Evensen JF, Berg K, Western A, Rimington C 1987 Photosensitizing efficiencies, tumor and cellular uptake of different photosensitizing drugs relevant for photodynamic therapy of cancer. Photochem Photobiol 46:713–722

Moreno G, Salet C 1985 Cytotoxic effects following micro-irradiation of cultured cells sensitized with hematoporphyrin derivative. Int J Radiat Biol 47:383–386

Paquette B, Ali H, Langlois R, van Lier J 1988 Biological activities of phthalocyanines. VIII. Photochem Photobiol 47:215–220

Sandberg S, Romslo I 1981 Phototoxicity of protoporphyrin as related to its subcellular localization in mice livers after short-term feeding with griseofulvin. Biochem J 198:67–74

Sandberg S, Romslo I 1982 Phototoxicity of uroporphyrin related to its subcellular localization in rat livers after feeding with hexachlorobenzene. Photobiochem Photobiophys 4:95–106

Santus R, Kohen C, Reyftman JP, Morliere P, Dubertret L, Tocci PM 1983 Permeation of lysosomal membranes in the course of photosensitization with methylene blue and hematoporphyrin: study by cellular microspectrofluorimetry. Photochem Photobiol 38:71–77

Schlegel RA, Phelps BM, Waggoner A, Terada L, Williamson P 1980 Binding of merocyanine 540 to normal and leukemic erythroid cells. Cell 20:321–328

Septimus M, Seiffert W, Zimmermann HW 1983 Uber hydrophobe Acridinfarbstoffe zur Fluorochromierung von Mitochondrien in lebenden Zellen. Histochemistry 79:443–456

Winkelman J 1961 Intracellular localization of hematoporphyrin in transplanted tumors. J Natl Cancer Inst 27:1369–1377

DISCUSSION

Dougherty: Protoporphyrin is known to enter cells. It is very hydrophobic, binds to LDL, and would be expected to be effectively retained. Yet in patients this drug is readily washed out. Why is this?

Moan: Highly lipophilic drugs such as protoporphyrin rapidly cross the cell membrane, but a fraction remains in the membrane. When that fraction is washed away, it will be replaced by intracellular dye.

Dougherty: Does the tetraphenylporphyrin monosulphonate, which is more hydrophobic, wash out as readily as protoporphyrin?

Moan: Apparently not quite as readily. The side groups may introduce binding to structures inside the cell.

Jori: Why should there be an equilibrium between protoporphyrin in different compartments of the cells? Could protoporphyrin be hydrated to haematoporphyrin in the cell? That might explain why it's cleared so fast.

Bonnett: It's conceivable, but not very likely.

Brown: Protoporphyrin is a substrate for ferrochelatase, which is present in many cells. It could be converted into haem.

Kessel: You could distinguish that readily because protoporphyrin fluoresces at 635 nm and haematoporphyrin fluoresces at 620 nm. We have never seen any evidence of protoporphyrin being transformed into haematoporphyrin.

The medium used to wash cells is important. Serum-free medium will not wash protoporphyrin out of cells; serum must be present and there is competition between the serum lipoproteins and the cells for the protoporphyrin.

Jori: Why aren't the dimers or trimers washed out of the cells?

Kessel: For the dimers and trimers the fluorescence is harder to see because the fluorescence yield is much lower.

Moan: We can see dimers by fluorescence methods; in HpD the dimers have about half the fluorescence quantum yield of the monomers (Moan & Sommer 1983).

Brown: It might be interesting to study protoporphyrin as it leaves cells, rather than simply looking at its disappearance from cells.

Jori: If protoporphyrin is converted into haem, would it be expected to leak out of the cell?

Bonnett: I don't think hydration of protoporphyrin happens. There is a low concentration of protoporphyrin and a high concentration of protohaem in the body but, as far as I am aware, haematoporphyrin and haematohaem do not occur naturally in cells.

Moan: Dr Kessel, your data on the washing of monomers of HpD from cells and the retention of the other components are similar to ours. It's hard to understand why protoporphyrin is washed away while other drugs with a similar lipophilicity are not.

Kessel: This may have something to do with molecular mass.

Dougherty: After injection of Photofrin II, protoporphyrin is not washed out of tumour tissue. When we extracted the tumours we found that protoporphyrin remained within the tissue cells whereas the monomers did not.

Henderson: We found the same when we extracted cells isolated from tumours.

Dougherty: Why does the monosulphonated phthalocyanine undergo endocytosis?

Moan: This drug aggregates extensively in solution and compounds that aggregate tend to be endocytosed.

Dougherty: Doesn't it disaggregate under the conditions in which you deliver the drug?

Moan: No, not to any large extent in serum, but as it enters the cells some of it seems to disaggregate.

Jori: Serum contains lipoproteins which might mediate the endocytosis.

Moan: Endocytosis is indicated by the presence of drug in the lysosomes. Exposure to light leads to disruption of the lysosomes, resulting in increased fluorescence of the drug.

We tried to study this in an intact tumour because Dr S. Bown's group (personal communication) has shown that phthalocyanines seem to be photodegraded *in vivo*, which would lead to a decrease in fluorescence rather than an increase. We injected a nude mouse carrying a tumour on one foot with tetrasulphonated aluminium phthalocyanine ($AlPcS_4$) and studied the fluorescence spectrum while irradiating the tumour via a fibre coupled to a dye laser. The fluorescence decreased, in agreement with Dr Bown's finding. Thus, in the tumour the drug does not localize in lysosomes, in contrast to what happens in our cells in culture. In the tumour I think $AlPcS_4$ localizes in extracellular structures.

Henderson: People often ask 'what is the primary cellular target of PDT?'. Would you agree that there is no single primary target?

Moan: Although many targets are affected, none seems to be responsible for photoinactivation under all conditions. This is true for DNA, for the machinery involved in DNA synthesis, for the plasma membrane and for the lysosomes.

Henderson: The primary target depends entirely on where the photosensitizer is localized at a particular time under given conditions.

Berenbaum: One of your basic assumptions was that the site of damage gives information about the localization of the sensitizer. In a small highly integrated object, such as a cell, that is a very unsafe assumption. If an electric light goes out, it doesn't necessarily mean that the site of damage is the lamp; it could be two miles away in a power station or it could be the switch on the wall. For example, cells might show inhibited DNA synthesis simply because the supply of substrates has been diminished by cytoplasmic damage.

Moan: By using low temperatures we can put [^{3}H]thymidine into the cells without it being incorporated into the DNA. Then the same amount is in all cells. Under such conditions we observe that PDT leads to a strong reduction of the DNA synthesis.

Berenbaum: But that's only one substrate of many. Also, couldn't physically damaged DNA be caused by singlet oxygen diffusing in from the cytoplasm?

Moan: From its short lifetime in cells ($< 1\ \mu s$), it is estimated that singlet oxygen cannot move more than 0.1 μm.

Berenbaum: If DNA synthesis is occurring, as you say, mainly around the nuclear membrane, that is near enough.

Moan: Less than 0.1 μm is very short. The cell is 10–20 μm in diameter. However, I agree with you that this approach gives only indirect evidence.

Hönigsmann: Is the DNA damage similar to that caused by UV irradiation—pyridimine dimer formation within the DNA?

Moan: The damage occurs mainly at guanine, which is oxidized. This leads to either a break or an alkali-labile site.

Bonnett: Is this a singlet oxygen reaction with guanine?

Moan: Yes, that has been demonstrated.

Hönigsmann: Presumably it's not a mutagenic lesion, because we don't see skin tumours in porphyric patients.

Moan: Many people, including my group and that of C. J. Gomer, have looked for mutagenesis induced by porphyrins and light in cells but have found no evidence for it.

Henderson: Nancy Oleinick says that the mutagenicity of phthalocyanines is greater than that of porphyrins (personal communication). The risks are lower than with ionizing radiation, but equal to those with UVC irradiation (Evans et al 1989).

Moan: But Ben-Hur et al (1987) came to different conclusions when using the same drug, chloroaluminium phthalocyanine.

Henderson: Dr Oleinick said that it depends on the method used to determine mutagenicity; it varies according to cell type.

Moan: She used two different cell lines: the LY-R and the LY-S strain of L5178 mouse lymphoma cells (10th International Congress on Photobiology, Jerusalem 1988, Abstract book, p 52).

Lim: Knowing the intracellular localizations of the various dyes, might we design one that is taken up primarily by tumour cells? For example, rhodamine 123 is taken up and retained selectively by mitochondria in some carcinoma cells (Bernal et al 1983).

Moan: Some positively charged dyes are thought to be selective for carcinoma cells in suspension. However, they are concentrated in the mitochondria of cancer cells by completely different mechanisms than those for the negatively charged dyes.

Dougherty: This presumably arises because of a charge differential between the inside and outside of cells.

Moan: Positively charged dyes frequently bind to DNA.

Jori: Powers (1988) claims that the binding is only to mitochondria.

Kessel: The 'multi-drug resistant' cells that develop resistance to a collection of drugs present a problem. Many common tumour cells are inherently multi-drug resistant and do not take up rhodamine 123.

Henderson: But do they take up porphyrins or other sensitizers?

Kessel: Yes; porphyrins seem to get into any malignant cell, and also many normal cell types. But if you make a cell resistant to adriamycin, it also fails to take up rhodamine 123.

Wilson: We've seen porphyrin resistance in multi-drug resistant Chinese hamster ovary cells. Measuring cell survival versus drug concentration at a fixed light dose, we found about 30% resistance.

Kessel: Whereas there may be 100- to 1000-fold resistance to the anthracyclines.

Henderson: But among ten unrelated cell lines some will take up more sensitizer than others; this effect may be totally coincidental. In our laboratory we have worked with cells that were 20- or 30-fold resistant to adriamycin. They showed the same uptake of sensitizer and the same amount of cell death as non-resistant cells.

References

Ben-Hur E, Fujihara T, Suzuki F, Elkind MM 1987 Genetic toxicology of the photosensitization of Chinese hamster cells by phthalocyanines. Photochem Photobiol 45:227–230

Bernal SD, Lampidis TJ, McIsaac RM, Chen LB 1983 Anticarcinoma activity of rhodamine 123, a mitochondrial specific dye. Science (Wash DC) 222:169–171

Evans HH, Rerko RM, Mencl J, Clay ME, Antunez R, Oleinick NL 1989 Cytotoxic and mutagenic effects of the photodynamic action of chloroaluminium phthalocyanine and visible light in L5178Y cells. Photochem Photobiol 49:43–47

Moan J, Sommer S 1983 Uptake of the components of hematoporphyrin derivative by cells and tumours. Cancer Lett 21:167–174

Powers SK 1988 Cationic dyes with mitochondrial specificity for phototherapy of malignant tumours. In: Neckers DC (ed) New directions in photodynamic therapy. Proc International Society for Optical Engineers (SPIE) 847:74–89

Tissue localization of photosensitizers and the mechanism of photodynamic tissue destruction

Barbara W. Henderson and David A. Bellnier*

Division of Radiation Biology, Roswell Park Memorial Institute, Buffalo, NY 14263, USA and **The Oncologic Foundation of Buffalo, Buffalo, NY 14203, USA*

Abstract. This paper outlines our present knowledge of photosensitizer tissue distribution, derived from preclinical animal studies, and relates it to the observed biological response to photodynamic therapy (PDT). Emphasis is placed on porphyrins (haematoporphyrin derivative (HpD), Photofrin II) and phthalocyanines (aluminium phthalocyanine sulphonate AlPcS). In mice, both groups of sensitizers show multiphasic plasma clearance kinetics with an initial rapid decline followed by further slow reduction. Residual amounts of Photofrin II are detectable 75 days after injection. Drug elimination occurs through urine and faeces, but faecal elimination predominates for Photofrin II. Circulating sensitizer greatly influences the mouse ear-swelling response, but not the foot response. Tumours and normal skin can be destroyed by vascular damage, if illumination occurs at times of maximal plasma sensitizer concentration, with no detectable sensitizer accumulation in tumour cells. Organ retention for both photosensitizer groups is similar and persistent. Organs rich in reticuloendothelial elements (liver, kidney, spleen) accumulate and retain the highest levels, skin and muscle the lowest, while normal brain tissue excludes sensitizer. The adrenal and pancreatic glands, as well as urinary bladder, also retain high amounts of Photofrin II. Tumour/skin ratios of 1 to 3:1 and 2 to 7:1 have been reported for porphyrins and sulphonated phthalocyanines respectively. Tissue destruction upon light exposure is not always correlated with photosensitizer levels, as is exemplified by liver and pancreas. Stromal sensitizer localization usually predominates in tumour and normal tissue, and often determines tumour response. Certain compounds, such as monosulphonated tetraphenylporphyrin and AlPcS, may favour parenchymal localization. The formed blood elements remain free of photosensitizer, while mast cells and macrophages accumulate especially large amounts and, upon illumination, release an array of vasoactive inflammatory and immune mediators.

1989 Photosensitizing Compounds: their Chemistry, Biology and Clinical Use. Wiley, Chichester (Ciba Foundation Symposium 146) p 112–130

Photodynamic therapy (PDT) of solid tumours usually involves systemic administration of tumour-localizing photosensitizing compounds and their subsequent local activation by light (Dougherty 1988). The presence of molecular oxygen within the treated tissue is a prerequisite for biological effectiveness, because cytotoxicity and eventual tumour destruction are mediated by the interaction between the triplet excited state of the sensitizer, oxygen and a biological substrate, generally through the formation of highly reactive singlet oxygen (Gomer & Razum 1984, Moan & Sommer 1985).

PDT produces rapidly developing coagulation necrosis of tumour cells, usually accompanied by severe effects on the microvasculature that are expressed as vasoconstriction/dilatation, platelet aggregation, microcirculatory stasis, endothelial destruction, and haemorrhaging (Bugelski et al 1981, Selman et al 1984, Star et al 1986). Normal tissue capillaries show microscopic damage similar to that in tumour vessels (Chaudhuri et al 1986, Fingar & Henderson 1987, Tseng et al 1988), which in skin may lead to such gross symptoms as oedema, erythema, and possibly necrosis (Dougherty 1988). These observations suggest that photosensitizers are also distributed to structures other than tumour and that photosensitization of these structures may contribute to the overall tissue response to PDT.

In this paper we shall review what is known about photosensitizer distribution among and within various tissues, and attempt to relate this distribution to the response to PDT.

Pharmacokinetics of photosensitizer distribution

The kinetics of photosensitizer distribution and elimination have been studied in mice for the porphyrin compound Photofrin II (PII), presently the only clinically used photosensitizer, and for chloroaluminium phthalocyanine sulphonate (AlPcS), a new agent for PDT undergoing extensive studies in animal and *in vitro* systems.

The plasma clearance kinetics after i.p. and i.v. injection of [^{14}C]Photofrin II, which is identical to Photofrin II in its physical and biological characteristics (Ho et al 1988), are shown in Fig. 1. After rapid absorption from the peritoneal cavity ($t_{1/2} = 1$ h), plasma clearance was best fit by a triexponential equation with elimination half lives of 4 h, 9 days and 36 days. At 24 h post-injection, about 1% of the total injected material could be detected in the circulation. This decreased to about 0.04% at 75 days. Intravenous drug administration showed different kinetics only in the more rapid initial clearance rate. More than 65% of the administered [^{14}C]PII was excreted over 192 h and 91% of the recovered material was found in the faeces.

Similar pharmacokinetic data have been reported for AlPcS by Weintraub et al (1988): 1% of injected material remained in the serum at 16 h and 0.05% at 96 h post-injection. Urinary excretion of AlPcS was predominant during the first 48 h after injection. Over a period of 30 days, however, similar amounts were excreted through faeces and urine.

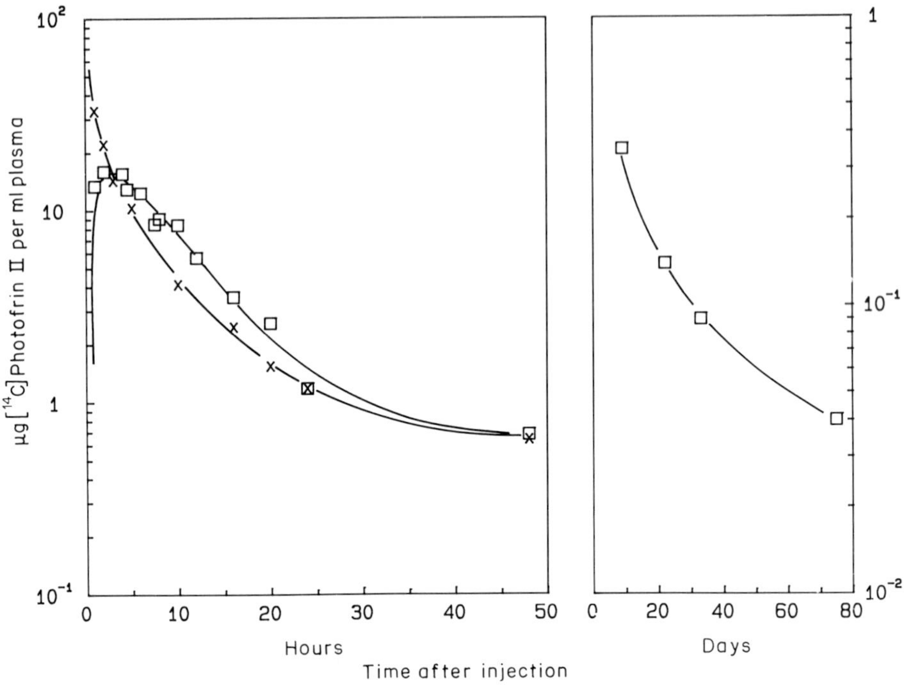

FIG. 1. Clearance of [^{14}C]Photofrin II from the plasma of DBA/2 Ha mice following i.p. (□) or i.v. (×) administration of 5 mg/kg. Each point is the mean of 3 to 6 animals; standard deviations no greater than 30%. Reprinted, with permission, from Bellnier et al (1989).

Although one might speculate that the persistence of Photofrin II in the circulation is responsible for the prolonged skin photosensitivity observed after use of this drug, data concerning the significance of circulating photosensitizer for the PDT response are scarce. In mice it was found that the ear-swelling response following PDT with PII directly correlated with circulating sensitizer levels, whereas it was independent of sensitizer levels in the exsanguinated ear tissue (Fig. 2). A similar relationship was observed for the expression of vascular PDT damage in the mouse back skin. In contrast, the mouse foot-swelling response is known to persist undiminished for at least one week after the injection, which is consistent with the prolonged tissue retention of Photofrin II.

The possible impact of circulating photosensitizer on PDT tumour response has been studied in mice carrying the RIF tumour (Table 1). Intracardiac injection of PII, followed immediately by tumour illumination for 30 min, produced complete tumour and normal tissue necrosis in the majority of animals at one-fifth of the dose necessary for the same response when tumours are illuminated 24 h after injection. Isolation of tumour cells 30 min post-injection

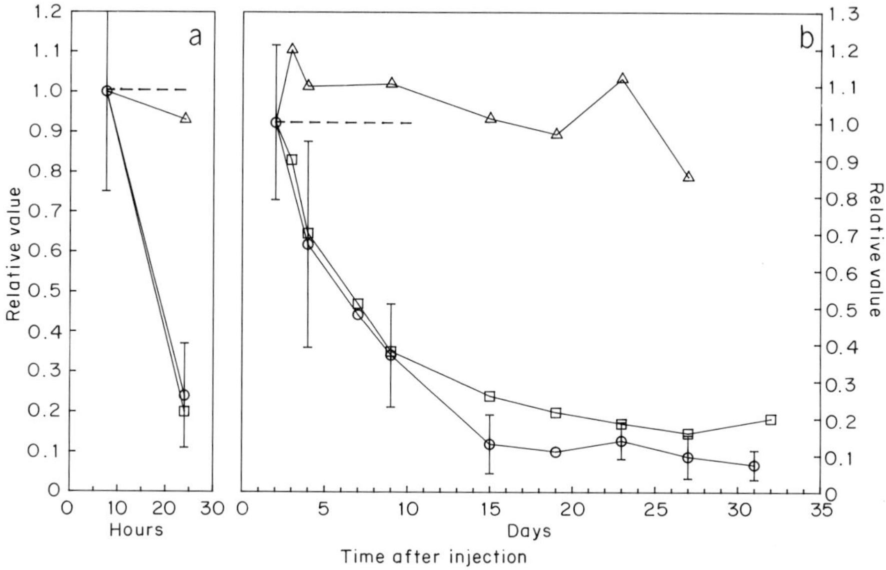

FIG. 2. Changes in the ear-swelling response (○) and foot response (----) to PDT, as well as in Photofrin II concentration in plasma (□) and exsanguinated ear tissue (△), as a function of time after photosensitizer administration. Values are normalized to the first data point in each panel to simplify ordinates. Ear-swelling response in Ros:(ICR) mice given 5 mg/kg Photofrin II and (a) 35 J/cm^2 and (b) 88.6 J/cm^2 of 630 nm light. Foot response in DBA/2 Ha mice given 4.2 mg/kg Photofrin II and 283.5 J/cm^2 of 630 nm light. Concentrations of Photofrin II in plasma and ear tissue of Ros:(ICR) mice were determined using [^{14}C]Photofrin II and are from pooled samples obtained from 3 to 6 mice/point. Reprinted, with permission, from Bellnier & Dougherty (1989).

and subsequent light exposure *in vitro* resulted in 100% cell survival, indicating that the tumour response achieved immediately after sensitizer injection was of a purely vascular nature. Similar effects can be achieved with protoporphyrin, but a five times higher dose of the very hydrophilic uroporphyrin produced no effect whatsoever. This is consistent with the observation by Nelson et al (1986) that uroporphyrin, although an efficient singlet oxygen producer, is ineffective in producing biological photodynamic damage, probably because it does not bind to a potential biological target. Some binding of sensitizer to vascular structures thus appears to be a prerequisite of vascular photosensitization.

In summary, in preclinical animal studies circulating photosensitizer can significantly contribute to the PDT response, but this contribution may depend greatly on the tissue being treated, as well as on the sensitizer used.

TABLE 1 RIF tumour response following PDT with circulating photosensitizer in mice

Photosensitizer	*Dose (mg/kg)*	*Vascular shutdown*	*Complete response Day 7*	*Cure Day 70*
Photofrin II	1.0	complete	8/10	2/10
	0.5	partial	0/10	0/10
	0.1	none	0/10	0/10
Uroporphyrin	5.0	none	none	none

Tumours were irradiated (135 J/cm^2, 75 mW/cm^2, 630 nm) immediately after intracardiac injection of sensitizer. Vascular shutdown in the treatment field determined by fluorescein angiography.

Photosensitizer distribution and retention in various organs

The distribution of photosensitizing porphyrins, in particular haematoporphyrin derivative (HpD), has been studied repeatedly (Gomer & Dougherty 1979, Kessel 1986). Figure 3 shows data on the distribution of [^{14}C]-labelled Photofrin II in mice after i.p. injection of 5 mg/kg. As with HpD, large amounts of porphyrin were found in the liver, kidney and spleen which are rich in reticuloendothelial components. Porphyrin accumulation was also high in two secretory organs, the adrenal and pancreatic glands, as well as the urinary bladder. The sensitizer appears to be almost totally excluded from normal brain. Differences in porphyrin uptake between transplantable, subcutaneous tumours in mice and tumour-surrounding skin are small: ratios (tumour/skin) of 1.08 (SMT-F mammary carcinoma, Fig. 3), 1.8 (SMT-F mammary carcinoma, Gomer & Dougherty 1979), and 2.2 (Lewis lung carcinoma, Kessel 1986) have been reported. Retention of Photofrin II in tissues is prolonged (Fig. 3). For example, at 75 days after injection, 16% of peak ^{14}C tissue levels were still found in the kidney, while the spleen had retained 61%. Nine percent of total injected porphyrin, as measured by ^{14}C activity, was found in the liver at that time.

Very similar distribution and retention patterns were observed in mice for AlPcS by Chan et al (1988) using fluorescence spectroscopy after drug extraction. Organ rankings of liver > spleen > small intestine > kidney > lung > heart > urinary bladder > stomach > colon > > > brain were found for photosensitizer concentration 12 days after injection. These authors also studied the ratios of drug uptake in tumour/skin in three different subcutaneously implanted mouse tumour lines and found these to vary from as little as approximately two (UV-2237 fibrosarcoma) to as much as seven (Colo 26 carcinoma).

Judgement of organ sensitivity to PDT and resulting toxicity of treatment to the animal based on photosensitizer distribution data can be problematic. For example, in our studies direct PDT treatment to the mouse liver, the organ accumulating and retaining the highest photosensitizer levels, carried out 24 h after injection of 5 mg/kg Photofrin II and applying up to 113 J/cm^2 (630 nm

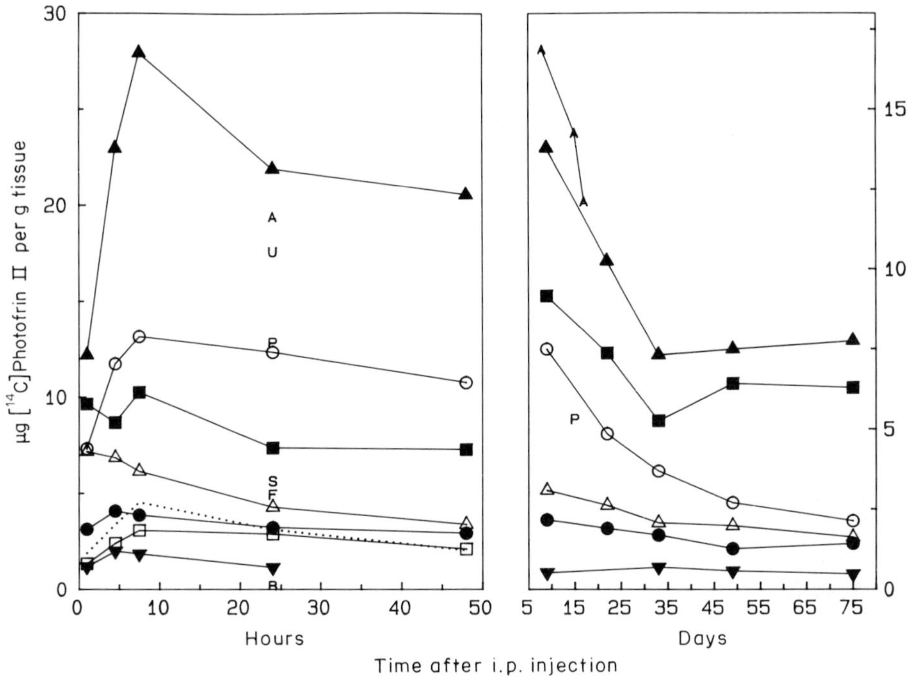

FIG. 3. Distribution of radioactivity in DBA/2 Ha mouse tissues. Animals received 5 mg/kg [^{14}C]Photofrin II i.p. and were killed at the indicated intervals after drug administration. Data are the averages for 3 to 6 animals, which differed by less than 30%. Concentrations in liver (▲), kidney (○), spleen (■), lung (△), heart (●), skin contralateral to tumour (□), muscle (▼), and SMT-F tumour (······) were calculated as if all radioactivity was accounted for by [^{14}C]Photofrin II. Letters indicate 24 h points for the following organs: A, adrenal gland; P, pancreas; S, stomach; F, femur; U, urinary bladder; B, brain. Reprinted, with permission, from Bellnier et al (1989).

light, 59.1 mW/cm^2, 7 mm diameter spot), caused only superficial tissue damage and no apparent ill effects to the animals. This lack of damage, despite high photosensitizer levels, can probably be attributed to the poor light penetration in liver tissue.

The pancreas presents a more puzzling phenomenon. As reported by Mang & Wieman (1987), normal murine pancreatic tissue accumulates porphyrin to the same extent as those of intrapancreatic tumours. However, the normal pancreas appears to be resistant to PDT photodestruction, while tumours are susceptible to damage. The low photobleaching rates of the photosensitizer during illumination of the normal pancreas point to possible differences in the photochemical and photophysical 'environment' of the pancreatic tissue as one reason for this uncharacteristic resistance to PDT.

Finally, Star et al (1986) reported almost equal tissue necrosis in rat subcutis isolated and treated in observation chambers and rat ears treated in air, despite the fact that there was a sixfold difference in HpD content between the tissues. The authors concluded that the biological effectiveness of PDT with HpD is not necessarily proportional to the total tissue porphyrin concentration.

Photosensitizer distribution in stromal and parenchymal tissue components

The strong affinity of photosensitizing porphyrins to stromal tissue components has long been noted. Bugelski et al (1981), studying the tissue distribution of HpD by autoradiographic techniques, observed grain density distributions 24 h after photosensitizer administration with ratios of 5:1 in favour of connective tissue elements in mouse liver, spleen, skin, stomach and SMT-F mammary carcinoma. Structures with high HpD accumulation included the capsules of liver and spleen, pseudocapsule and connective tissue septa in the tumour, serosa and submucosa of the stomach and the dermis. Likewise, the highest AlPcS fluorescence has been observed in the serosa and submucosa of the normal rat colon (Barr et al 1988) as well as in tumour stroma. The exact localization of photosensitizer within the stroma is unclear—binding to the fibrous tissue matrix (collagen, elastic and reticular fibre) has been suggested (Musser et al 1982, Straight & Spikes 1985), and a study showing the early destruction of fibrous elements in the subendothelial zone of capillaries after PDT using Photofrin II, chlorin e_6 aspartate and AlPcS tends to support this suggestion (Nelson et al 1988). Endothelial photosensitizer localization has long been assumed, because of the frequently observed endothelial PDT damage, and Weintraub et al (1988) have found that AlPcS is localized in sinusoidal endothelial cells of spleen and lymph nodes.

By far the most prominent accumulation of photosensitizers, porphyrins as well as phthalocyanines, occurs in the components of the reticuloendothelial system, which include the free and fixed macrophages of connective and tumour tissues, Kupffer cells in the liver, and red pulp macrophages of the spleen (Bugelski et al 1981, Chan et al 1988). Similarly, mast cells have been observed to retain high levels of HpD (P. J. Bugelski, unpublished data 1981).

The apparent lack of photosensitizer uptake and retention by the formed blood elements, despite their direct access to the highest sensitizer concentrations, is surprising. We could detect no [^{14}C]-labelled Photofrin II in mouse red and white blood cells after injection of 5 mg/kg of sensitizer. Similarly, Chan et al (1988), using flow cytometric analysis of cell suspensions from dispersed tumours after *in vivo* sensitizer administration, found lymphocytes and polymorphonuclear cells to be mostly devoid of AlPcS.

Despite the apparent predominance of stromal photosensitizer distribution, it is clearly proven that tumour cells can take up and strongly bind a number of different photosensitizers. In the stroma-poor RIF mouse fibrosarcoma, we found at least 90% of the total tumour [^{14}C]Photofrin II content to be firmly attached to tumour cells after enzymic tumour dispersion (Henderson & Fingar 1989). Likewise, Chan et al (1988) found that tumour cells were contained in the most highly fluorescent cell fractions after *in vivo* AlPcS administration and subsequent dispersion of the mouse Colo 26 carcinoma.

The factors that govern photosensitizer distribution among the various tissue components have not yet been fully elucidated, but the binding characteristics of the photosensitizer to plasma proteins may be a determining influence (Jori et al 1984). Kessel et al (1987), examining the tissue distribution of various sulphonated derivatives of tetraphenylporphyrin in a subcutaneously implanted pancreatic mouse tumour, found that those analogues (one or two adjacent sulphonates) which bound preferentially to plasma lipoproteins partitioned into neoplastic cells, whereas analogues binding to albumin (three or four sulphonates) accumulated in tumour stroma. Curiously, despite these differences, the overall tumour accumulation was reported to be the same for all compounds tested.

In view of the heavy stromal accumulation of photosensitizers, particularly in subcutaneously transplantable rodent tumours, it is not surprising that PDT effects on the tumour bed greatly influence tumour response. This situation is clearly reflected in the events leading to tumour destruction of the RIF fibrosarcoma in mice by Photofrin II-mediated PDT. At minimum curative PDT doses, where sensitizer uptake in tumour cells is too low to cause direct tumour cell inactivation, tumour cell kill is facilitated by tumour bed effects leading to vascular disruption and nutritional/oxygen deprivation (Henderson et al 1985, Fingar & Henderson 1987, Fingar et al 1988, Henderson & Fingar 1989). At high doses of Photofrin II, where the high sensitizer accumulation would allow extensive direct photodestruction of tumour cells, the rapidly ensuing tumour hypoxia resulting from vascular damage limits this possibility by limiting the oxygen available for the photodynamic process. Eventual tumour cell death is therefore again a consequence of destruction of the tumour bed.

Subcutaneous rodent tumour models, including the tumour-surrounding skin, seem to be particularly prone to vascular disruption by PDT, whereas tumours of more highly differentiated type may be somewhat more resistant (Chapman et al 1988). However, some degree of vascular damage in PDT-treated tumours and normal tissue has been observed in practically all preclinical and clinical studies. For treatment selectivity it might therefore be desirable to shift the mechanism of tumour destruction away from vascular (i.e. tumour bed) effects towards direct parenchymal effects. To this end we are studying the relative abilities of different photosensitizers to induce these diverse effects using two experimental designs (Fig. 4). Assessment of reduction in clonogenic tumour cell survival after sensitizer accumulation *in vivo*, cell isolation and light exposure *in vitro* provides information about the tumour cell-localizing ability and efficiency of light activation of the photosensitizer, while excluding possible

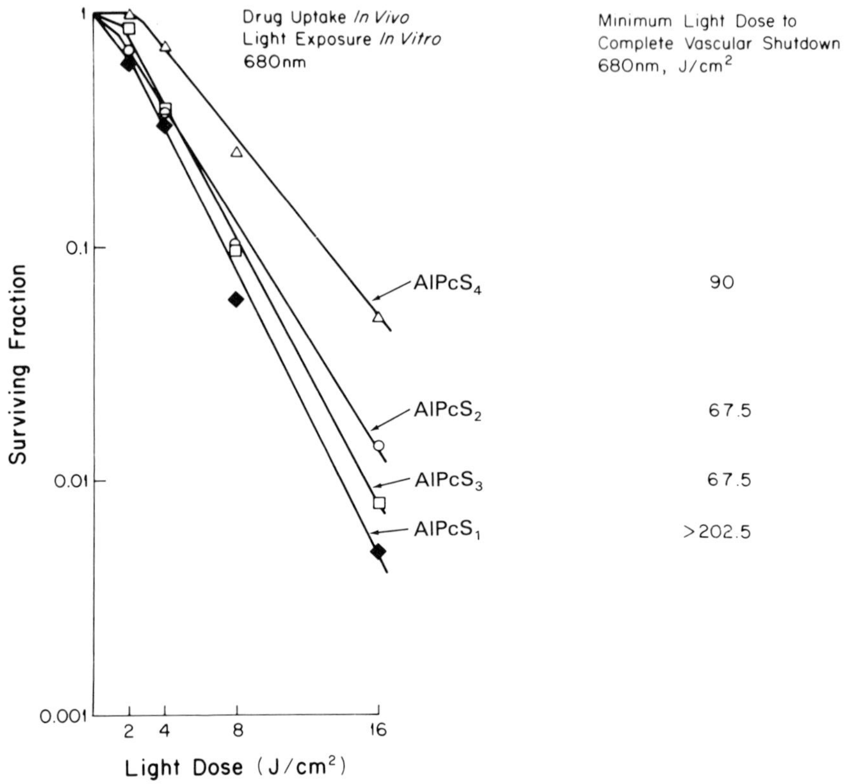

FIG. 4. Parenchymal and stromal effects in C3H/HeJ mice following PDT with sulphonated phthalocyanines of different degrees of sulphonation. Parenchymal effects were assessed as the lethality to RIF tumour cells of 680 nm light *in vitro* after sensitizer exposure (6.8 μmol/kg) for 24 h *in vivo*. Cells were pooled from at least two tumours. Stromal effects in the back skin (7 mm diameter field) were assessed in parallel by fluorescein angiography (10 mg/kg fluorescein i.v. immediately after PDT, visualization of dye distribution under UV light).

light and oxygen limitations which might be encountered during *in vivo* light treatment. The use of fluorescein angiography to establish a minimum dose for vascular shutdown in the skin (exclusion of fluorescein injected after PDT treatment from the treatment field) allows quantitation of vascular photosensitivity. It appears that some sensitizers, for example monosulphonated phthalocyanine and possibly monosulphonated tetraphenylporphyrin, have superior potential for tumour cell inactivation while being relatively sparing to the vasculature. Whether these characteristics are sufficient to improve PDT treatment selectivity remains to be seen.

The participation of mast cells in PDT-induced skin phototoxicity has been well studied. Kerdel et al (1987) described mast cell degranulation, accompanied

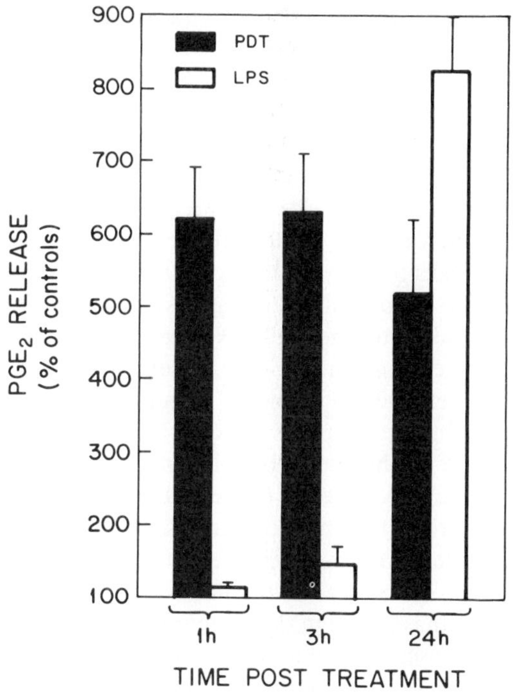

FIG. 5. PGE_2 release, assessed by an $[^{125}I]PGE_2$ radio immunoassay from mouse peritoneal macrophages in response to PDT or lipopolysaccharide (LPS, *Escherichia coli*) treatment. Cells were exposed to 25 µg/ml Photofrin II for 1 h, followed by 1 J/cm^2 of 630 nm light, or to 10 µg/ml LPS continuously. Medium was harvested at 1, 3 and 24 h. Values are the mean ± SD from duplicate plates and duplicate determinations.

by increased circulating levels of histamine and leukotriene C_4, during the early phase of the ear-swelling response in mice following HpD-mediated PDT. The release of various arachidonic acid metabolites from rat serosal mast cells under *in vitro* conditions following protoporphyrin-mediated, but not uroporphyrin-mediated, PDT has also been reported (Lim et al 1986).

Finally, we have observed the release of prostaglandin E_2 from mouse peritoneal macrophages following PDT with Photofrin II (Fig. 5), indicating that these cells, which exhibit extremely high affinity for the photosensitizer, may also significantly contribute to the overall PDT response. In particular, they may play a role in the occasionally observed shock-like lethality in mice after high PDT doses. This phenomenon was found to be diminished after administration of prostaglandin synthesis inhibitors (Gomer et al 1988). Also, prostaglandin release may contribute to the PDT-induced immunosuppressive effects which have been observed in mice, such as the inhibition of contact hypersensitivity (Elmets & Bowen 1986) or the transient decrease of splenic natural killer cells (Gomer et al 1988).

Conclusions

Although tissue accumulation of those effective photosensitizers which have been studied in detail in rodent models usually leads to a therapeutic differential between tumour and surrounding normal tissue, the photosensitizers are far from being exclusively selective for tumour cells. Furthermore, it is safe to say that some contribution to the overall PDT response can be elicited from all those structures which bind and retain the photosensitizer, although this contribution may not always be commensurate with sensitizer levels. Circulating photosensitizer may be reflected in studies of skin photosensitivity, particularly if the rodent ear-swelling response is used as a model. The predominant localization and retention of photosensitizers in connective tissue elements causes normal tissue and tumour bed effects, expressed as vascular photosensitivity, which in the case of poorly differentiated, subcutaneously transplanted rodent tumours may become the governing factor in tumour destruction. Among the connective tissue elements, mast cells and macrophages may be involved in the induction of vascular damage, and in the systemic toxicity sometimes observed in mice after high PDT doses, by shedding an array of vasoactive inflammatory and immune mediators. Certain photosensitizers may favour parenchymal tumour localization. The reasons for this are being investigated and include solubility characteristics, binding properties to plasma proteins, and possible receptor-mediated cellular uptake. It is hoped that further drug development studies will place importance on the search for such compounds. Tumour destruction can be achieved in animal models by exploiting circulating photosensitizer levels. This can be shown with porphyrins, and also with other compounds such as mono-L-aspartyl chlorin e_6. The latter acts as an efficient photosensitizer if irradiation is performed shortly after administration while blood levels are high, but all photodynamic activity is lost by 24 h post-injection because of the drug's rapid *in vivo* clearance (Gomer & Ferrario 1988). Although these pharmacokinetic characteristics provide the advantage of avoiding prolonged skin photosensitivity, tumour destruction is achieved at the expense of treatment selectivity. On the other hand, a number of cationic, photosensitizing dyes appear to possess true tumour cell selectivity, but their sensitizing efficiency is too low to make them practical alternatives at this time (Oseroff et al 1986, 1988).

Finally, very little is known about photosensitizer pharmacokinetics and distribution in man. Based on *in vivo* fluorescence measurements and biological response to PDT, tumour-selective Photofrin II accumulation in man seems to be far superior to that observed in rodent models, with tumour:skin ratios of the order of 8:1, 48 h after drug injection (T. Mang, personal communication

1989). In the human bladder, on the other hand, treatment selectivity is very poor, despite apparently good porphyrin fluorescence differential between tumour and normal tissue (Benson et al 1982). This lack of selectivity is reminiscent of the high porphyrin levels found in normal rodent bladder tissue. From the scarce, anecdotal information available concerning human porphyrin-induced skin photosensitivity, it appears that its prolonged time-course corresponds most closely with the rodent foot response model, rather than the ear-swelling response.

Further photosensitizer development will certainly be guided by the information obtained from preclinical animal experiments such as those described here. Only clinical experience will determine whether this will translate into improved therapeutic effects of PDT.

Acknowledgement

This investigation was supported in part by PHS Grant Number CA 42278, awarded by the National Cancer Institute, DHHS.

References

Barr H, Tralau CJ, MacRobert AJ et al 1988 Fluorescence photometric techniques for determination of microscopic tissue distribution of phthalocyanine photosensitizers for photodynamic therapy. Lasers Med Sci 3:81–86

Bellnier DA, Dougherty TJ 1989 The time course of cutaneous porphyrin photosensitization in the murine ear. Photochem Photobiol 49:369–373

Bellnier DA, Ho Y-K, Pandey RK, Missert JR, Dougherty TJ 1989 Distribution and elimination of Photofrin II in mice. Photochem Photobiol 50:221–228

Benson RC Jr, Farrow GM, Kinsey JH et al 1982 Detection and localization of in situ carcinoma of the bladder with hematoporphyrin derivative. Mayo Clin Proc 57:548–555

Bugelski PJ, Porter CW, Dougherty TJ 1981 Autoradiographic distribution of hematoporphyrin derivative in normal and tumor tissue of the mouse. Cancer Res 41:4606–4612

Chan W-S, Marshall JF, Lam GYF et al 1988 Tissue uptake, distribution, and potency of the photoactivatable dye chloroaluminium sulfonated phthalocyanine in mice bearing transplantable tumors. Cancer Res 48:3040–3044

Chapman JD, Chetner MP, Arnfield M et al 1988 Exploitation of PDT-induced hypoxia in rat tumors to produce cures. Lasers Med Sci (abstract issue, July) abstract 80

Chaudhuri K, Goldblatt PJ, Kreimer-Birnbaum M et al 1986 Histological study of the effect of hematoporphyrin derivative photodynamic therapy on the rat jejunum. Cancer Res 46:2950–2953

Dougherty TJ 1988 Photodynamic therapy. In: Withers HR, Peters LJ (eds) Medical radiology—innovations in radiation oncology. Springer-Verlag, Berlin & Heidelberg, p 175–188

Elmets CA, Bowen KD 1986 Immunological suppression in mice treated with hematoporphyrin derivative photoradiation. Cancer Res 46:1608–1611

Fingar VH, Henderson BW 1987 Drug and light dose dependence of photodynamic therapy: a study of tumor and normal tissue response. Photochem Photobiol 46:837–841

Fingar VH, Mang TS, Henderson BW 1988 Modification of photodynamic therapy-induced hypoxia by Fluosol-DA (20%) and carbogen breathing in mice. Cancer Res 48:3350–3354

Gomer CJ, Dougherty TJ 1979 Determination of [^{3}H] and [^{14}C] hematoporphyrin derivative distribution in malignant and normal tissue. Cancer Res 39:146–151

Gomer CJ, Ferrario A 1988 Preclinical evaluation of mono-L-aspartyl chlorin e6 (NPe6) in photodynamic therapy. Lasers Med Sci (abstract issue, July) abstract 155

Gomer CJ, Razum NJ 1984 Acute skin response in albino mice following porphyrin photosensitization under oxic and anoxic conditions. Photochem Photobiol 40:435

Gomer CJ, Ferrario A, Hayashi N et al 1988 Molecular, cellular and tissue responses following photodynamic therapy. Lasers Surg Med 8:450–463

Henderson BW, Fingar VF 1989 Oxygen limitation of direct tumor cell kill during photodynamic treatment of a murine tumor model. Photochem Photobiol 49:299–304

Henderson BW, Waldow SM, Mang TS, Potter WR et al 1985 Tumor destruction and kinetics of tumor cell death in two experimental mouse tumors following photodynamic therapy. Cancer Res 45:572

Ho Y-K, Pandey RK, Missert JR et al 1988 Carbon-14 labeling and biological activity of the tumor-localizing derivative of hematoporphyrin. Photochem Photobiol 48:445–449

Jori G, Beltrami M, Reddi E et al 1984 Evidence for a major role of plasma lipoproteins as hematoporphyrin carriers in vivo. Cancer Lett 24:291–297

Kerdel FA, Soter NA, Lim HW 1987 In vivo mediator release and degranulation of mast cells in hematoporphyrin derivative-induced phototoxicity in mice. J Invest Dermatol 88:277–280

Kessel D 1986 Porphyrin–lipoprotein association as a factor in porphyrin localization. Cancer Lett 33:183–188

Kessel D, Thompson P, Saatio K et al 1987 Tumor localization and photosensitization by sulfonated derivatives of tetraphenylporphine. Photochem Photobiol 45:787–790

Lim HW, Parker D, Marcus AJ 1986 Generation of eicosanoids from mast cells exposed to protoporphyrin and irradiation. Clin Res 34(2):763A

Mang TS, Wieman TJ 1987 Photodynamic therapy in the treatment of pancreatic carcinoma: Dihematoporphyrin ether uptake and photobleaching kinetics. Photochem Photobiol 46:853–858

Moan J, Sommer S 1985 Oxygen dependence of the photosensitizing effect of hematoporphyrin derivative in NHIK-3025 cells. Cancer Res 45:1608–1610

Musser DA, Wagner JM, Datta-Gupta N 1982 The interaction of tumor localizing porphyrins with collagen and elastin. Res Commun Chem Pathol Pharmacol 36:251–259

Nelson JS, Sun C-H C, Berns MW 1986 Study of the in vivo and in vitro photosensitizing capabilities of Uroporphyrin I compared to Photofrin II. Lasers Surg Med 6:131–136

Nelson JS, Liaw L-H, Orenstein A et al 1988 Mechanism of tumor destruction following photodynamic therapy with hematoporphyrin derivative, chlorin, and phthalocyanine. J Natl Cancer Inst 80:1599–1605

Oseroff AR, Ohuoha D, Ara G et al 1986 Intramitochondrial dyes allow selective in vitro photolysis of carcinoma cells. Proc Natl Acad Sci USA 83:9729–9733

Oseroff AR, Wadwa K, Smith S et al 1988 Cationic photosensitizers permit selective photochemotherapy of carcinomas and melanomas in vitro and in vivo. Lasers Med Sci (abstracts issue, July), abstract 156

Selman SH, Kreimer-Birnbaum M, Klaunig JE et al 1984 Blood flow in transplantable bladder tumors treated with hematoporphyrin derivative and light. Cancer Res 44:1924–1927

Star WM, Marijnissen HPA, van den Berg-Blok AE et al 1986 Destruction of rat mammary tumor and normal tissue microcirculation by hematoporphyrin derivative photoradiation observed in vivo in sandwich observation chambers. Cancer Res 46:2532–2540

Straight RC, Spikes JD 1985 Preliminary studies with implanted polyvinyl alcohol sponges as a model for studying the role of neointerstitial and neovascular compartments of tumors in the localization, retention and photodynamic effects of photosensitizers. Adv Exp Med Biol 193:77–89

Tseng MT, Reed MWR, Ackerman DM et al 1988 Photodynamic therapy induced ultrastructural alterations in microvasculature of the rat cremaster muscle. Photochem Photobiol 5:675–681

Weintraub H, Abramovici A, Altman A et al 1988 Toxicity, tissue distribution and excretion studies of aluminium phthalocyanine tetrasulfonate in normal mice. Lasers Life Sci 2:185–196

DISCUSSION

Dougherty: With Photofrin II, and apparently most of the aluminium phthalocyanine derivatives, the most important initial effect of photodynamic therapy (PDT) on tumours is vascular damage. How does that happen?

Henderson: The photosensitizer is probably bound to endothelial cells, and is probably also in the subendothelial matrix. When these endothelial cells are irradiated, other cells (except for the blood cells, apparently) release various vasoactive substances, including clotting factors, which set in motion vascular damage. Platelet aggregation occurs first and then the microvasculature is blocked and stasis follows, which results in endothelial damage. Therefore the endothelium may be damaged initially or later, as a consequence of stasis. I don't think it's one particular effect.

Bown: Stasis doesn't make the calibre of the vessels reduce, which is what one sees.

Henderson: That's why I think other factors are involved.

Dougherty: Is vasodilatation the first effect?

Henderson: No; vasoconstriction comes first. For example, in the prostaglandin cascade the precursors of prostaglandin E (PGE) are important vasoconstrictors. They are converted into prostaglandins E and I which are vasodilators.

Lim: Although umbilical cord endothelial cells usually produce predominantly PGI_2, when treated with porphyrin and irradiation they selectively release PGF_2, which is a vasoconstrictor. That could be one factor that causes the immediate vasoconstriction.

Henderson: Have you looked at leukotrienes?

Lim: No. However, we examined the effects of porphyrins on mast cells and on cutaneous eicosanoid metabolism. Porphyrins and radiation markedly affect the generation of eicosanoids (Lim, this volume).

Dougherty: Is this effect inhibited by indomethacin or aspirin?

Lim: The release of PGD_2 by mast cells and the release of PGF_2 by endothelial cells are both inhibited by aspirin.

Greaves: Dr Henderson, you described PGE_2 release from mouse peritoneal macrophages in response to PDT and lipopolysaccharide treatment (Fig. 5). The time-courses were different. Was there any difference in the sensitivity to indomethacin of those two methods of release? With different time-courses you can have either cyclooxygenase-mediated production or non-enzymic release of prostaglandins.

Henderson: I haven't compared the effect of the inhibitors on the two forms of PGE_2 release. The porphyrin-induced release is completely blocked by indomethacin, almost completely blocked by meclofenamate, and partly blocked by dexamethasone, if I administer the inhibitor 24 h before PDT. The release is not significantly influenced by hydrocortisone and quinacrine. The inhibitors seem to affect both the phospholipase and the cyclooxygenase systems. I observed a curious effect with calcium; Ca^{2+} depletion completely blocks the release of PGE_2 induced by freeze–thawing but has no effect on the PDT-induced release.

Lim: That is true with mast cells also; PGD_2 release is Ca^{2+} independent.

Dougherty: What is this telling us?

Henderson: It is probably related to the mitochondrial Ca^{2+} pool. When the cells are frozen, all the membranes are disrupted and all the Ca^{2+} is available for chelation with EDTA. In PDT, if only the outer membrane is disrupted and the mitochondrial Ca^{2+} pool remains and slowly releases Ca^{2+} in response to PDT, this intracellular pool can provide the Ca^{2+} to activate phospholipase.

Lim: I am intrigued by the difference between the effect of circulating Photofrin II on the foot pad and on the ear (Fig. 2). Could this arise because the concentration of mast cells is much higher in the ear than in the foot pad? Might the ear-swelling response be correlated with plasma levels of photosensitizer because the combination of porphyrin and light would lead to complement activation, resulting in generation of anaphylatoxin and degranulation of mast cells? A different mechanism which correlates with tissue levels of photosensitizer might operate in the foot pad.

Henderson: We have wondered about that.

Dougherty: The location of histamines would answer this. You could determine that with antihistamines.

Lim: Yes.

Henderson: But with antihistamines Dr Lim found very small decreases in the ear swelling.

Lim: The antihistamine studies were done in guinea pigs. In mast cell-deficient mice the phototoxic response, as measured by ear thickness, is suppressed compared to that in mast cell-sufficient animals. What happened with the back skin?

Henderson: The response in the back skin looks almost identical to that in the ear, in terms of the minimum light dose required to cause vascular shutdown. The minimum dose increases rapidly with decreasing plasma photosensitizer levels. In contrast, the foot response is the same whether you irradiate after one or ten days.

Lim: Is the swelling on the foot pad immediate? There is a significant increase in ear thickness at the completion of irradiation.

Dougherty: It would be detectable immediately in the foot but it takes several hours to develop fully.

Lim: That's true for the ear; the swelling reaches a maximum 12 hours after exposure to light.

Wilson: On the issue of why phthalocyanines do not show their full potential *in vivo*, is it too simplistic to suggest that this arises because the predominant effect of PDT with both Photofrin II and the phthalocyanines is vascular damage but it's much less efficient with phthalocyanines than with Photofrin II?

Henderson: We can predict what the efficiency of direct tumour cell inactivation by PDT with phthalocyanines should be in our model. From the survival curves after *in vivo* drug uptake and *in vitro* irradiation we can extrapolate how much light it would take to kill a given number of cells. We have measured the light transmission through skin and tumour. Therefore we know how much light is available to tumour cells during treatment.

Wilson: I'm suggesting that the two or three logs of kill on the *in vitro* assay may be irrelevant in terms of tumour response. Many logs of kill from the vascular effect may be required and phthalocyanines do not seem to produce very efficient vascular damage.

Henderson: But with the phthalocyanines in our model only about 10 J is required to reduce cell survival by four logs. You are getting close to having a direct effect if the vessels stay open.

Dougherty: Is there self-shielding by the photosensitizer here?

Henderson: Yes, we measure light transmission to correct for that.

Brown: In your study of the clearance of Photofrin II in mice you find monomers in the faeces. Thus at some stage the aggregates may convert into monomers. We studied Photofrin II in rat bile (Fig. 1, p 128) and found that its fluorescence peaks about 30 min after injection. When we hydrolyse the drug in bile and remeasure the fluorescence, there is initially no increase. Because aggregated Photofrin II does not fluoresce as strongly as monomers, this suggests the presence of only monomers. After 60 min the amount of fluorescence in the hydrolysed fraction is higher than that in the unhydrolysed sample, suggesting that rat bile contains a significant proportion of aggregates at that time. We confirmed this by HPLC on various fractions. We can also use ^{14}C-labelled monomers of haematoporphyrin within the whole Photofrin II mixture, or selectively label an aggregate fraction. Our HPLC results show that maximum haematoporphyrin excretion occurs later when we inject Photofrin II than when

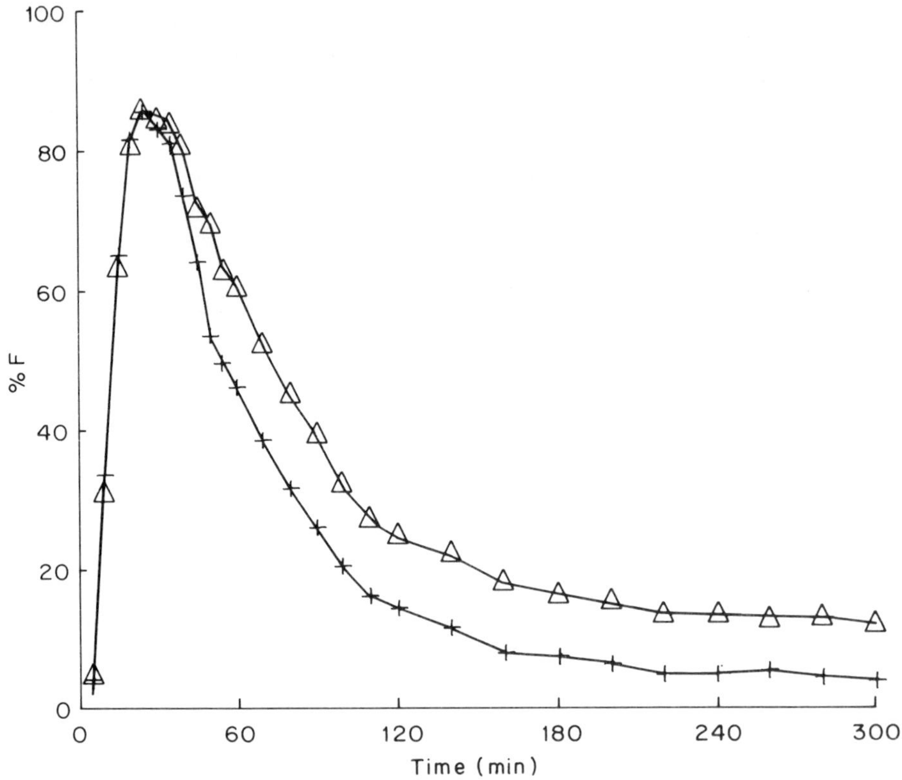

FIG. 1. (Brown). Fluorescence of rat bile after polyhaematoporphyrin (Photofrin II) administration. The curves show fluorescence before (+) and after (△) acid hydrolysis, which converts aggregates into monomeric haematoporphyrin. Initially there is little increase in fluorescence upon hydrolysis, implying that the porphyrin in bile is almost completely monomeric. Later there is a progressive increase in fluorescence after hydrolysis, indicating that an increasing fraction of the drug is present as aggregate.

we use a sample of 'pure' haematoporphyrin. Thus in bile there is some conversion of aggregate into monomers. There may be even more at a later stage before the drug is excreted in faeces.

Bown: We have also found the normal pancreas to be very resistant to PDT. Could there be something in the pancreas that's mopping up singlet oxygen?

Henderson: T.S. Mang and colleagues (personal communication) are cannulating the lymphatics to see if there is some factor that protects against photodamage. They believe that the porphyrin fluorescence in the pancreas is localized in the lymphatic channels.

Kessel: Does the light penetrate into the pancreas?

Bown: Yes; we can shine light through the pancreas and damage the bowel beyond it. Schroder et al (1988) found the same.

Dougherty: If you implant a tumour in the pancreas which has the same concentration of Photofrin II as the normal pancreas, PDT destroys the tumour and not the pancreas.

Kessel: The fact that you see no porphyrins in red blood cells, lymphocytes and platelets is interesting. Do these cells have LDL receptors?

Jori: No.

Dougherty: Do patients with protoporphyria have fragile red blood cells?

Lim: They have elevated levels of protoporphyrin in younger red blood cells. As the red cell ages, the fluorescence decreases. It is thought that the protoporphyrin leaks out of the cells. But some protoporphyrin is present and when the cells are irradiated membrane damage and photohaemolysis are induced.

Dougherty: That seems very odd, because earlier we heard that protoporphyrin is not retained in cells (Moan et al, this volume).

Greaves: In erythropoietic protoporphyria (EPP) the porphyrins are synthesized constitutively in the red blood cells. They don't get there from outside.

Lim: In protoporphyria there is a defect in ferrochelatase. The substrate for this enzyme is protoporphyrin.

Dougherty: Perhaps protoporphyrin is continually produced.

Brown: No, I don't think so.

Lim: Protoporphyrin is produced in the bone marrow and as the red blood cells come out a certain number contain elevated protoporphyrin. As the cells age many of them no longer fluoresce because the protoporphyrin leaks out of the cells.

Dougherty: So the protoporphyrin does eventually clear.

Lim: Yes.

Henderson: How about uroporphyrin in uroporphyria?

Lim: It is primarily in the plasma; porphyrin levels in the red blood cells are usually not elevated in patients with porphyria cutanea tarda.

Henderson: Does that create photosensitivity?

Dougherty: Yes, which is strange.

Henderson: If you put uroporphyrin into the plasma *in vivo* there are no phototoxic effects when the animal is exposed to light.

Moan: Uroporphyrin is also produced by fibroblasts in the skin.

Jori: So that is the origin of the cutaneous photosensitivity.

Hönigsmann: A major cause of skin photosensitivity is the porphyrin serum level. High levels of protoporphyrin are also found in erythrocytes, but not in the serum, in people who are intoxicated with lead. These people are not photosensitive, in contrast to protoporphyric patients. The weak binding mechanism within the erythrocyte in EPP allows leakage of protoporphyrin, resulting in photosensitivity.

Moan: In EPP, when the erythrocytes pass through the skin capillaries and are exposed to light, protoporphyrin is released because the binding sites to haemoglobin are destroyed (A. Brün & S. Sandberg, personal communication). The protoporphyrin passes from the membranes of the red cells to the walls of the narrow vessels when the red blood cells are squeezed through the vessels in the skin. We layered EPP red cells onto fibroblasts growing in tissue culture and irradiated them with red light. Protoporphyrin was transferred down to the fibroblasts without injury to the red cells (A. Brün, S Sandberg & J. Moan, unpublished data).

References

Lim HW 1989 Effects of porphyrins on skin. In: Photosensitizing compounds; their chemistry, biology and clinical use. Wiley, Chichester (Ciba Found Symp 146) p 148–158

Moan J, Berg K, Kvam E et al 1989 Intracellular localization of photosensitizers. In: Photosensitizing compounds: their chemistry, biology and clinical use. Wiley, Chichester (Ciba Found Symp 146) p 95–111

Schroder T, Wen Chu I, Sperling M, Bell RH, Brackett K, Joffe SN 1988 HpD uptake and PDT in pancreatic carcinoma. J Surg Oncol 38:4–9

Immunosuppression in phototherapy

Stephen E. Ullrich, Joseph Alcalay, Lee Ann Applegate and Margaret L. Kripke

Department of Immunology (178), The University of Texas, M. D. Anderson Cancer Center, 1515 Holcombe Blvd, Houston, TX 77030, USA

Abstract. The successful use of phototherapy, especially psoralen plus UVA (PUVA) therapy, in the treatment of a variety of skin diseases is well known. Because the pathology of diseases such as vitiligo, alopecia and lichen planus is thought to involve immune mechanisms, the beneficial effect of PUVA may be due to immunosuppression. PUVA treatment can induce suppression in two ways. In the first (local suppression) psoralen is applied topically, the skin is irradiated with UVA and the contact allergen is applied directly to the irradiated skin. The induction of contact hypersensitivity (CHS) is suppressed and suppressor cells are found in the spleens of treated animals. Systemic suppression results from the injection of psoralen followed by exposure to UVA. The contact allergen is then applied at a distant non-irradiated site. CHS is suppressed and antigen-specific suppressor cells are found in the spleens of treated mice. The ability to induce specific immunosuppression may provide novel methods of inhibiting unwanted immune responses. We have demonstrated that graft rejection and the induction of graft-versus-host disease can be suppressed in an antigen-specific manner by UV radiation. Thus phototherapy may provide promising new treatments for suppressing graft rejection and perhaps may be beneficial in the treatment of autoimmune disease and allergic reactions.

1989 Photosensitizing Compounds: their Chemistry, Biology and Clinical Use. Wiley, Chichester (Ciba Foundation Symposium 146) p 131–147

Combined therapy using photosensitizing compounds and non-ionizing radiation has been used to treat a variety of diseases. Psoralen plus UVA radiation (PUVA) is used to treat various skin disorders, including psoriasis, vitiligo, mycosis fungoides, alopecia areata and lichen planus (Morison et al 1979). Edelson et al (1987) reported that extracorporeal photophoresis (ingestion of psoralen followed by *ex vivo* exposure of peripheral blood lymphocytes to UVA) was successful in treating disseminated cutaneous T cell leukaemia. Studies with experimental animals have demonstrated that PUVA therapy can be used to prevent the rejection of skin allografts (Granstein et al 1987). Photodynamic therapy using haematoporphyrin derivatives and long wavelength light has been used in the diagnosis and treatment of cancer (Dougherty 1987).

The mechanisms underlying the successful treatment of these disorders are probably as diverse as are the diseases themselves. The mechanism of tumour cell killing in photodynamic therapy is thought to involve the production of singlet oxygen by the excited photosensitizer which results in the irreversible oxidation of vital cellular components. One report, however, suggests that the tumour cell killing may be secondary to the destruction of collagen and connective tissue in the subendothelial zone of the tumour capillary wall (Nelson et al 1988). The inhibition of DNA synthesis by PUVA may play an important role in the management of psoriasis, a disease characterized by marked epidermal cell proliferation. The mechanism involved in the management of other diseases, such as vitiligo, alopecia and lichen planus, which are not characterized by hyperproliferation, is not clear. It has been suggested that the pathology of these diseases involves immune mechanisms and that the beneficial effect of PUVA therapy results from immunosuppression. This paper reviews the mechanisms involved in the generation of immunosuppression during phototherapy.

Local suppression of the immune response by photosensitizing compounds

The term local suppression generally refers to experiments in which a contact allergen is applied directly to PUVA-treated skin. These experiments were modelled after those of Toews et al (1980) who exposed mice to low doses (400 J/m^2) of UVB radiation and sensitized the animals through the irradiated skin. They found that the ability to induce contact hypersensitivity (CHS) was significantly suppressed. The depression of CHS appears to be related to the depletion of Langerhans cells at the site of sensitization by the UV radiation. Subsequent studies demonstrated the existence of T suppressor cells in the spleens of these mice (Elmets et al 1983). In a similar manner, Horio & Okamoto (1983) were able to suppress CHS in guinea pigs with PUVA treatment. A topical preparation of 8-methoxypsoralen (8-MOP) was applied to the skin and the skin was exposed to UVA radiation. Three days later, a contact allergen was applied to the PUVA-treated skin. Contact hypersensitivity was significantly suppressed in these animals. The suppression of CHS in PUVA-treated guinea pigs may also be related to the depletion of Langerhans cells by PUVA treatment (Okamoto & Horio 1981). The suppression is hapten specific because resensitization of the PUVA-treated animals with a different contact allergen resulted in a response that was indistinguishable from that found in the controls. Because the suppression was reversed by cyclophosphamide, Okamoto & Horio suggested that suppressor cells may be involved.

Results from our laboratory have confirmed that suppressor cells are induced after topical psoralen and UVA treatment. Two different psoralens were used in this study—8-MOP, a bifunctional psoralen, and angelicin, a monofunctional psoralen. Monofunctional psoralens are generally much less phototoxic and have much less mutagenic activity. We found that the results were similar regardless of the type of psoralen used. Treatment with 8-MOP and UVA resulted in severe gross phototoxicity (erythema and eroded skin), depletion of Langerhans cells and depression of CHS. While there was no overt phototoxicity when angelicin and UVA were used, there was a significant depletion of Langerhans cells and CHS was depressed. Cell transfer studies indicated that suppressor cells were present in the spleens of the 8-MOP + UVA-treated and the angelicin + UVA-treated animals. From these studies we conclude that: (1) Local PUVA treatment does induce suppressor cells; (2) Suppression and suppressor cell induction occur in the absence of any overt phototoxicity, indicating that the inflammation and phototoxicity that normally result from PUVA treatment are not a necessary condition for the induction of suppressor cells.

Systemic suppression of the immune response by PUVA

The term systemic suppression was also initially used in studies describing the effects of UVB on the immune response. Mice exposed to a large dose (30 to 40 kJ/m^2) of UVB radiation and sensitized with a contact allergen at a distant non-irradiated site have a depressed CHS response (Noonan et al 1981). Kripke et al (1983) showed that intraperitoneal (i.p.) injection of 8-MOP followed by exposure to UVA (18 to 36 kJ/m^2) also resulted in a systemic suppression of CHS. A delay of three to five days between PUVA treatment and sensitization was required to induce suppression. Antigen-specific suppressor T cells were present in the spleens of the PUVA-treated mice. Like the suppressor cells induced by UVB radiation, PUVA-induced suppressor cells were active in suppressing the induction of CHS but not the elicitation of the response. However, the phenotypes of the two suppressor cell populations were different; the UVB-induced suppressor cells were Lyt 1^+, 2^-, whereas the PUVA-induced suppressor cells were Lyt 1^+, 2^+ (Ullrich & Kripke 1984).

A fascinating aspect of the systemic suppression of immunity by UVB and/or PUVA is the antigenic specificity of the suppressor cells. Perhaps one of the most interesting applications of this procedure is the suppression of the immune response to alloantigens by UV radiation. Exposure of mice to UV radiation followed by subcutaneous injection of alloantigen results in the suppression of delayed hypersensitivity (DTH) to the alloantigen. Spleen cells from the UV-treated antigen-sensitized mice were unable to proliferate to the alloantigen in the *in vitro* mixed lymphocyte reaction. Associated with the suppression is the induction of antigen-specific suppressor T cells (Ullrich 1986). Allograft survival was significantly prolonged in mice treated with UV radiation and sensitized with alloantigen. The suppressive effect was specific in that BALB/c mice exposed to UV and injected with C3H spleen cells would accept C3H allografts but reject C57Bl/6 grafts. Also, exposure to UV radiation coupled with alloantigenic sensitization can be used to suppress the induction of graft-versus-host disease. Acute graft-versus-host disease can be induced in mice by injecting mature spleen cells and bone marrow cells into lethally X-irradiated allogeneic recipients. The injected T cells recognize the antigens of the host as foreign and the reaction of these cells against the host is characterized by wasting, weight loss, alopecia, diarrhoea and, eventually, death. When spleen cells from mice exposed to UV and injected with alloantigen were used to reconstitute the lethally X-irradiated recipients, there was a significant prolongation of survival of these animals compared to controls that received normal spleen cells (Ullrich & Magee 1988). Thus it is possible to use UV exposure coupled with antigenic sensitization to suppress graft rejection in an antigen-specific manner. Exposure to UV radiation may also be used to suppress other 'unwanted' immune reactions. Experiments by Hauser et al (1984), demonstrating inhibition of experimental allergic encephalitis in mice by UV radiation, suggest that phototherapy may be useful in suppressing autoimmune disease.

TABLE 1 Comparison between local and systemic suppression induced by PUVA

	Local	*Systemic*
Dose of UVA	Low	High
Site of sensitization	Irradiated skin	Distant unexposed skin
Depletion of Langerhans cells	Yes	No
Induction of suppressor cells	Yes	Yes
Specificity of suppressor cells	?	Yes

Role of soluble factors in the induction of suppression after UV exposure

The mechanism by which UV exposure causes the induction and/or activation of splenic suppressor cells is not entirely clear. The penetrating power of UV radiation is obviously not sufficient for the spleen to be irradiated. On the basis of observations that urocanic acid is the major UV-absorbing component of the stratum corneum and that urocanic acid undergoes a *trans* to *cis* isomerization whose action spectrum (peak activity at 270 nm) closely follows that described for the induction of systemic suppression of CHS, DeFabo & Noonan (1983) suggested that *cis*-urocanic acid may be the photoreceptor involved in the induction of suppression by UV radiation. Data supporting this hypothesis come from studies of Ross et al (1986). Sensitization of mice with herpes simplex virus through *cis*-urocanic acid-treated skin can suppress DTH. In addition, the intravenous injection of *cis*-urocanic acid can result in systemic suppression of the response against the virus (Ross et al 1988). Furthermore, Noonan et al (1988) injected *cis*-urocanic acid into mice and found that splenic antigen-presenting cell function is suppressed in a manner similar to that observed after total body exposure to radiation (Noonan et al 1981). These findings support the suggestion by DeFabo & Noonan that urocanic acid is the photoreceptor for UV radiation.

On the other hand, results from our laboratory suggest that urocanic acid may not be the photoreceptor. In a way similar to that seen in mice, exposure of marsupials (*Monodelphis domestica*) to UV radiation can suppress CHS. The advantage of using marsupials is that, unlike mice, *M. domestica* possesses the photoreactivating repair enzyme that removes and repairs UV-induced pyrimidine dimers. Can the activation of the DNA repair mechanism by photoreactivating light overcome the UV-induced suppression of CHS? To answer this question, marsupials were exposed to UVB radiation and then irradiated with long wavelength photoreactivating light (300–500 nm). There was a 60 to 80% reversal of the suppression of CHS after exposure to photoreactivating light. Because activation of the DNA repair mechanism overcame the immunosuppressive effect of UV radiation, we conclude that DNA, and not urocanic acid, is the primary target of UV radiation.

Keratinocyte-derived immunosuppressive factors may also play a role in the induction of suppression by UV radiation. Keratinocytes make a variety of immunoregulatory substances, including interleukins (IL) 1 and 3, prostaglandins and colony-stimulating factors. Keratinocytes also produce immunosuppressive factors. Schwarz et al (1986) demonstrated that supernatants from UV-irradiated keratinocytes can suppress the induction of CHS in mice. An additional activity, inhibition of IL-1-induced thymocyte proliferation, has also been described for a 40 kDa protein present in the supernatants of UV-irradiated keratinocytes (Schwarz et al 1987). Whether the suppression of CHS is a function of this contra-IL-1 protein or is due to an additional inhibitory cytokine remains to be seen.

One characteristic of the suppression induced by UV radiation is its selectivity. Mice exposed to large doses of UV become susceptible to challenge with UV-induced tumours because of the induction of tumour-specific suppressor cells. However, most other immune responses in these animals, including allograft rejection, antibody production and lymphocyte proliferation in response to mitogens and antigens, are intact (Kripke 1981). Similarly, in mice exposed to a single dose of UV and sensitized with a contact allergen, only the response to that particular contact allergen is suppressed. If the systemic suppression induced by UV radiation can be attributed to the release of soluble factors by UV-irradiated keratinocytes, one would also expect the suppression induced by these factors to be selective. We attempted to address this question by examining the range of immune responses affected by supernatants from UV-irradiated keratinocytes. We were able to suppress DTH to alloantigen by injecting these supernatants but, contrary to the initial report of Schwarz (1986), supernatants from keratinocytes irradiated with UVB radiation were ineffective in suppressing CHS. On the other hand, treatment of the keratinocytes with UVA radiation (filtered to remove all contaminating UVB) produced a factor that could suppress CHS but not DTH. Thus it appears that different wavelengths of UV radiation can cause the production of different suppressive cytokines from keratinocytes.

These results suggest that the suppression of CHS and DTH by UV radiation occurs by different mechanisms.

How these data relate to the *in vivo* suppression of CHS and DTH by UV radiation is not entirely clear. Exposure of mice to FS-40 sunlamps, which emit 65% of their energy in the UVB range and 35% in the UVA region of the spectrum, suppresses both DTH and CHS. One might argue that the UVA present in the radiation from the FS-40 bulbs is suppressing CHS and the UVB component is suppressing DTH. In agreement with this hypothesis are the findings of Lynch et al (1981), who demonstrated that exposure of mice to broad-band UVA radiation could suppress CHS. However, experiments by DeFabo & Noonan (1983) using monochromatic light demonstrated that the optimal wavelength for suppressing CHS was within the UVB region of the spectrum (270 nm). In addition, we cannot suppress CHS with large doses of UVA radiation (filtered to remove all contaminating UVB radiation), suggesting that the UVA-induced suppressive factor is not involved in the *in vivo* suppression of CHS by UV radiation.

An alternative explanation for the differences between the suppression that results from injection of factors from UV-irradiated keratinocytes and the suppression of DTH and CHS in mice after exposure to the FS-40 sunlamps involves the production of anti-suppressive factors. Besides producing the factors that suppress DTH and CHS, keratinocytes might also produce factors that selectively inhibit the suppressive activity of one but not both factors. The production of the anti-suppressive factor would depend upon the wavelength of light used to irradiate the keratinocytes. Exposure to UVA would produce both suppressive substances and a factor that inhibits the expression of the substance that suppresses DTH. The net result would be suppression of CHS and not DTH. Similarly, exposure to UVB would produce both suppressive substances and a factor that inhibits the expression of the substance that inhibits CHS, the net result being suppression of DTH. We suggest that these anti-suppressive factors are either short-range or degraded *in vivo*, which would explain the difference between our *in vivo* and *in vitro* results.

Suppressive factors have also been discovered in the serum of UV-irradiated mice. Swartz (1984) and Harriott-Smith & Halliday (1986) found that the transfer of serum from UV-irradiated animals to normal recipients can suppress CHS in the recipient mice. Little is known at the present time about the identity of these factors. Suppression of CHS has also been achieved by injecting IL-1 into mice. Robertson et al (1987) found that injecting IL-1 into mice before sensitization with contact allergens could suppress the development of CHS. The suppression was sensitive to the effect of indomethacin, indicating a role for prostaglandin. These authors suggest that the inflammation caused by UV exposure results in the release of mediators such as IL-1 and prostaglandin that eventually suppress the immune response. Whereas Robertson et al (1987) could suppress CHS with IL-1, Harriott-Smith & Halliday (1986) were unable to detect

IL-1 in the serum from the UV-irradiated mice and Schwarz et al (1987) describe a contra-IL-1 activity in their suppressive supernatants. Furthermore, the addition of indomethacin did not abrogate the suppressive activity found in supernatants from UV-irradiated keratinocytes (Schwarz et al 1986). We have found that indomethacin does not inhibit production of the suppressive factor produced by UV-irradiated keratinocytes. Also, no urocanic acid was detectable in our supernatants (H. Morrison, personal communication) and while there was IL-1 in our suppressive supernatants there was an equivalent amount of IL-1 in the control supernatants from non-irradiated keratinocytes, suggesting that IL-1 was not responsible for the suppression observed in our experiments. Taken as a whole, these results suggest a role for soluble factors in the induction of suppression by UV radiation. Yet there remain a number of important issues that must be resolved. Is urocanic acid the photoreceptor or is the primary event DNA damage? How do the various suppressive substances interact with each other? Is there a cascade of events, do these factors work sequentially or does UV radiation cause the release of multiple factors that independently induce suppression? What is the biochemical composition of the various suppressive substances? How do these substances induce suppressor cells? Is the systemic suppression induced by PUVA the result of the production of suppressive substances by cells within the epidermis?

Conclusion

There are two main ways of inducing immunosuppression by PUVA treatment. In local suppression, the hapten is applied directly to PUVA-treated skin. In systemic suppression, mice are injected with psoralen, exposed to UVA and the hapten is applied at a distant non-irradiated site. In either case, suppressor cells are found in the spleens of the treated animals. The ability to induce antigen-specific suppressor cells by phototherapy may provide a useful method of suppressing undesirable immune reactions. For example, we have shown that graft rejection and the induction of graft-versus-host disease can be suppressed by the antigen-specific suppressor cells induced with UV radiation. Future uses of phototherapy may include the suppression of graft rejection and perhaps autoimmune disease and allergic reactions. While it is not clear how splenic suppressor cells are induced after irradiation of the skin, recent studies suggest that soluble suppressive factors are involved. Investigation of the biochemistry and mechanism of action of these suppressive factors should provide valuable insights into how suppressor cells are induced during phototherapy.

Acknowledgements

Research by the authors was supported by NIH grants R23-CA-42936 and RO1-CA-44713 and a grant from the Dermatology Foundation.

References

DeFabo EC, Noonan FP 1983 Mechanism of immune suppression by ultraviolet irradiation in vivo. I. Evidence for the existence of a unique photoreceptor in skin and its role in photoimmunology. J Exp Med 157:84–98

Dougherty TJ 1987 Photosensitizers: therapy and detection of malignant tumors. Photochem Photobiol 45:879–889

Edelson R, Berger C, Gasparro F et al 1987 Treatment of cutaneous T-cell lymphoma by extracorporeal photochemotherapy: preliminary results. N Engl J Med 316:297–303

Elmets CA, Bergstresser PR, Tigelaar RE, Wood PJ, Streilein JW 1983 Analysis of the mechanism of unresponsiveness produced by haptens painted on skin exposed to low dose UV radiation. J Exp Med 158:781–794

Granstein RD, Askari M, Whitaker D, Murphy GF 1987 Epidermal cells in the activation of suppressor lymphocytes: further characterization. J Immunol 138:4055–4062

Harriott-Smith TG, Halliday WJ 1986 Circulating suppressor factors in mice subjected to UV radiation and contact hypersensitivity. Immunology 57:207–211

Hauser SL, Wiener HL, Che M, Shapiro ME, Gilles F, Letvin NL 1984 Prevention of experimental allergic encephalomyelitis (EAE) in the SJL/J mouse by whole body ultraviolet irradiation. J Immunol 132:1276–1281

Horio T, Okamoto H 1983 Immunologic unresponsiveness induced by topical application of hapten to PUVA-treated skin in guinea pigs. J Invest Dermatol 80:90–93

Kripke ML 1981 Immunological mechanisms in ultraviolet radiation carcinogenesis. Adv Cancer Res 34:69–106

Kripke ML, Morison WL, Parrish JA 1983 Systemic suppression of contact hypersensitivity in mice by psoralen plus UVA radiation (PUVA). J Invest Dermatol 81:87–92

Lynch DH, Gurish MF, Daynes RA 1981 Relationships between epidermal Langerhans cell density ATPase activity and the induction of contact hypersensitivity. J Immunol 126:1892–1897

Morison WL, Parrish JA, Epstein JH 1979 Photoimmunology. Arch Dermatol 115:350–355

Nelson JS, Liaw L-H, Orenstein A, Roberts WG, Berns MW 1988 Mechanism of tumor destruction following photodynamic therapy with hematoporphyrin derivative, chlorin, and phthalocyanine. J Natl Cancer Inst 80:1599–1605

Noonan FP, Kripke ML, Pedersen GM, Greene MI 1981 Suppression of contact sensitivity by UV radiation is associated with defective antigen presentation. Immunology 43:527–533

Noonan FP, DeFabo EC, Morrison H 1988 cis-Urocanic acid, a product formed by ultraviolet B irradiation of the skin, initiates an antigen presentation defect in splenic cells in vivo. J Invest Dermatol 90:92–99

Okamoto H, Horio T 1981 The effect of 8-methoxypsoralen and long-wave ultraviolet light on Langerhans cells. J Invest Dermatol 77:345–346

Robertson B, Gahring L, Newton R, Daynes RA 1987 In vivo administration of IL-1 to normal mice decreases their capacity to elicit contact hypersensitivity responses: prostaglandins are involved in this modification of the immune response. J Invest Dermatol 88:380–387

Ross JA, Howie SEM, Norval M, Maingay BA, Simpson TJ 1986 Ultraviolet-irradiated urocanic acid suppresses delayed hypersensitivity to herpes simplex virus in mice. J Invest Dermatol 89:230–233

Ross JA, Howie SEM, Norval M, Maingay BA 1988 Systemic administration of urocanic acid generates suppression of the delayed type hypersensitivity response to herpes simplex virus in a murine model of infection. Photodermatology 5:9–14

Schwarz T, Urbanska A, Gschnait F, Luger TA 1986 Inhibition of the induction of contact hypersensitivity by a UV-mediated epidermal cytokine. J Invest Dermatol 87: 289–291
Schwarz T, Urbanska A, Gschnait F, Luger TA 1987 UV-irradiated epidermal cells produce a specific inhibitor of IL-1 activity. J Immunol 138:1457–1463
Swartz RP 1984 Role of UVB-induced serum factors in suppression of contact hypersensitivity in mice. J Invest Dermatol 83:305–307
Toews GB, Bergstresser PR, Streilein JW 1980 Epidermal Langerhans cell density determines whether contact hypersensitivity of unresponsiveness follows skin painting with DNFB. J Immunol 124:445–449
Ullrich SE, Kripke ML 1984 Mechanisms in the suppression of tumor rejection produced in mice by repeated UV irradiation. J Immunol 133:2786–2790
Ullrich SE 1986 Suppression of the immune response to allogeneic histocompatability antigens by a single exposure to UV radiation. Transplantation (Baltimore) 42:287–291
Ullrich SE, Magee M 1988 Specific suppression of allograft rejection by treatment of recipient mice with UV radiation and allogeneic spleen cells. Transplantation (Baltimore) 46:115–119

DISCUSSION

Dougherty: Is the formation of pyrimidine dimers in DNA lethal if it is not reversed?

Ullrich: Yes, if you irradiate cells in culture you can kill them.

Dougherty: Does this cause release of cytokines?

Ullrich: It may do. Unfortunately, we can't study the effect of DNA repair by photoreactivating light in the mouse keratinocyte system because mice lack the repair enzyme. UV irradiation probably results in a cascade of events and many cytokines may be involved. Ray Daynes found that recombinant IL-1 suppresses contact hypersensitivity (CHS) *in vivo*. That's a prostaglandin-mediated event (Robertson et al 1987). Tom Schwarz et al (1986) showed that indomethacin does not inhibit the production of the suppressive factor produced by UV-irradiated keratinocytes. Inflammation may or may not be involved, but after exposure to UVB or PUVA there is a fair amount of inflammation. Probably many things are happening and we may be looking at only a part of the cascade or at different parallel courses.

van den Bergh: How many cross-links are necessary to kill a cell?

Hönigsmann: One is sufficient if it is unrepaired.

Henderson: Are cross-links repaired in psoriatic skin?

Hönigsmann: Some people believe that they are repaired.

Jori: There may be reactivating enzymes acting on cross-links.

Hönigsmann: Perhaps there is a combination of post-replication repair and excision repair.

Moan: Has anyone compared treatment of psoriasis with UVA alone with treatment using the same dose of UVA plus a high dose of visible light? That might tell us something about the photoreactivating enzyme involved.

Hönigsmann: Bill Gange used 8-methoxypsoralen (8-MOP) and induced monofunctional adducts with wavelengths greater than 380 nm (Gold et al 1988). This caused persistent photosensitivity for more than two weeks. After that the induction of cross-links was much easier, which is the opposite of photoreactivation. Thus if you combine UVA and visible light you will obtain more cross-links.

Moan: Patients with psoriasis benefit from exposure to sunlight, which contains an enormous amount of visible, potentially photoreactivating, light.

Hönigsmann: We do not want to treat patients in an uncontrolled manner in the sun, because of possible mutagenic effects.

Jori: What is the evidence that urocanic acid is photoisomerized, Dr Ullrich?

Ullrich: trans-Urocanic acid was irradiated with UV and *cis* and *trans* urocanic acid were distinguished by HPLC. However, Henry Morrison couldn't find urocanic acid in our suppressive supernatants.

Lim: Could the difference between your results and those of Frances Noonan arise because she used mouse and you used possum? Does *Monodelphis domestica* have urocanic acid and, if so, can it isomerize *cis* to *trans*?

Ullrich: I don't know. Lee Applegate, using possums, could reverse the depletion of Langerhans cells caused by UVB radiation by irradiating with photoreactivating light (LeVee et al 1988). The Langerhans cells approached control levels, but 100% recovery was never seen.

Lim: How quickly does that occur?

Ullrich: They looked after 48 h—it's relatively quick.

Henderson: Gruner et al (1985) showed that PDT can prevent allograft rejection. They treated skin grafts before transplantation with haematoporphyrin derivative (HpD) and visible light. Does this work by affecting the Langerhans cells?

Ullrich: It has been suggested that irradiating the graft with UVB light before transplantation, which reduces the number of Ia-positive cells, reduces the immunogenicity of the graft. Rick Granstein transplanted the skin and then gave the animals PUVA therapy to induce suppression. Those grafts were also successful. We did it a little differently, trying to induce specific suppressor cells first and investigating what happens.

Moan: It has been debated whether PUVA treatment works exclusively by damaging DNA, or whether its action also involves membrane damage, notably for the monofunctional drugs such as psoralens.

Ullrich: Cells have 8-MOP receptors and 8-MOP binds to cell membranes. That may be another effect, but the studies are preliminary. I don't know whether the photoreactivating light would affect the 8-MOP bound to the cell membrane.

Moan: That may be part of the story. Local immunosuppression might involve damage of the cell membranes.

Henderson: Membrane damage does occur, and probably contributes to the efficiency of this therapy. Prostaglandin release, which is a membrane-related phenomenon, has been observed after treatment with ionizing (Rubin 1988) and UV radiation (DeLeo et al 1985), as well as PDT.

Ullrich: Craig Elmets suppressed CHS using haematoporphyrin and visible light. Membrane damage may be involved in that. There are probably many things going on, which may help to explain different results in different systems.

Jori: It has been proposed that singlet oxygen plays a role in the membrane damage generated by photoactivation of 8-MOP in cells. This would make quite a good parallel with red light PDT.

Ullrich: We tested a number of singlet oxygen scavengers; the results were inconclusive and suggest that it may not be just singlet oxygen. We can't totally overcome the suppressive effects in animals exposed to UV and injected with scavengers.

Bonnett: Is there any evidence that 8-MOP generates singlet oxygen and, if so, does it react with DNA to destroy guanine residues, for example?

Jori: De Mol et al (1981) reported that 8-MOP generates singlet oxygen. If there is no other quencher and if 8-MOP is not bound to DNA but is simply in solution, the singlet oxygen quantum yield may be very high. However, this may not be what happens in cells, although it is claimed that singlet oxygen is important in the observed membrane damage.

Bonnett: It seems strange that 8-MOP should produce singlet oxygen, because the triplet energy must be quite high—a long way above 94 kJ/mol. I wonder if the excited photosensitizer transfers an electron to oxygen, which would lead to a superoxide reaction producing hydrogen peroxide.

Jori: The lifetime of triplet 8-MOP is sufficient and singlet oxygen has been detected *in vitro* in solution.

Moan: Singlet oxygen luminescence produced by excitation of psoralens in cells has not been reported.

Henderson: The common mechanism for the induction of eicosanoid release from cells by ionizing radiation, UV and PDT is probably lipid peroxidation of the membrane. The cyclooxygenase and phospholipase systems are both sensitive to the amount of peroxide in cells.

Lim: Ray Daynes used slow-releasing long-acting indomethacin pellets implanted in the animal to reverse the systemic suppression by UVB (Chung et al 1986). That also suggests a role for prostaglandins, but they are not the only factor.

Greaves: There is much evidence *in vitro* and *in vivo* that UVB irradiation generates prostaglandins and indomethacin-sensitive inflammation, but there is no convincing evidence that PUVA therapy induces prostaglandin production in irradiated skin. In our *in vivo* studies in skin of the induction of acute inflammation by PUVA, we failed to show generation of prostaglandins, although we demonstrated *in vivo* induction of prostaglandin biosynthesis by UVB irradiation (Black et al 1980, Plummer et al 1978).

Hönigsmann: Dr Ullrich implied that there is no essential difference in the immunosuppression produced by UVB and by the phototoxic reaction of PUVA. You didn't UVB and PUVA irradiate *Monodelphis domestica* because PUVA lesions are probably not photoreactivated, unlike UVB lesions. It would be interesting to look for changes in the immune response after exposing the animal to only photoreactivating light.

Moan: Photoreactivation repairs only the pyridimine dimers.

Hönigsmann: It also acts on the UV-induced 6-4 photoproduct.

Moan: Yes, but that effect is secondary; when you remove the pyrimidine dimers more enzymes will be available to work on the 6-4 photoproduct.

Dougherty: In PUVA therapy of psoriasis, psoralen is administered orally and the patient is irradiated within a short period of time, presumably when there is a large amount of circulating drug. Therefore you would expect a large amount of vascular damage. The patient is exposed to the minimum effective dose. If you overdose the patient with light, do you induce skin necrosis?

Greaves: Yes.

Dougherty: Is there any selectivity between the psoriatic plaques and the normal tissue?

Greaves: There is a predilection for phototoxicity in normal skin, the psoriatic skin being protected.

Dougherty: At lower light doses you can eventually destroy the plaque and the normal tissue heals; that might be more to do with repair than with selective damage.

Greaves: Therapeutic selectivity seems to disappear when a certain UV dose is exceeded. Perhaps a different mechanism, one of tissue damage, operates at high UV doses in PUVA therapy.

Hönigsmann: This loss of selectivity and increased damage to normal skin rather than to psoriatic plaques is just an optical phenomenon relating to the light scattering in psoriatic lesions.

Dougherty: If you have high levels of circulating photosensitizer is the phototoxicity inherently non-selective?

Lim: If PUVA medication were given systemically, the phototoxic response would not be selective; if an overdose of light were given, both normal skin and plaque would be damaged.

Rook: There may be selectivity. Rapidly dividing cells appear to take up more psoralen than slowly dividing cells, as has been shown with activation of cells in the immune system.

Dougherty: What is the time interval between administering the drug and irradiation?

Hönigsmann: With the usual drug preparations, it's two hours. Light treatment takes 10 to 30 min.

Dougherty: How quickly does the drug clear?

Hönigsmann: There's no photosensitivity after eight hours. It takes about 24 hours to clear psoralens completely.

Moan: The rapidly dividing cells in psoriatic skin must have many LDL receptors because of their need for lipids to produce membranes. It might be possible to use a drug which is carried by LDL to these cells.

Hönigsmann: In East Germany, Diezel et al (1980) tried to use haematoporphyrin derivative to treat patients with psoriasis. They observed severe photosensitivity, particularly scarring of psoriatic areas. But this may simply reflect a problem with dosimetry.

Dougherty: At the University of California at Irvine a topical application of porphyrins was used, but because the vehicle didn't work very well the efficacy was uncertain. A few patients treated with porphyrins for malignant disease have also had psoriasis. Some people have been convinced that, with the proper dosimetry, porphyrin treatment of psoriasis could be feasible. But you can't do that with photosensitizing drugs such as Photofrin II.

Hönigsmann: It also depends on the light source.

Dougherty: The usual PUVA sources were used, but treatment was not optimized.

Rook: Could dimethyl sulphoxide (DMSO) be used as a vehicle for porphyrins?

Dougherty: That was tried. But another porphyrin has been reported to penetrate through some tissues. It might go through skin.

Jori: Psoralens can be applied topically using DMSO.

Greaves: The topical application of drugs in the treatment of psoriasis often exploits the fact that most active agents penetrate psoriatic plaques more readily than adjacent skin, thus conferring some selectivity on the treatment.

Jori: Is any general photosensitivity, which would imply that the drug enters the circulation, evident after topical application of 8-MOP?

Hönigsmann: Yes. After application to more than 30% of the skin surface one can find measurable amounts of 8-MOP in the serum.

Jori: Is this when DMSO is used as the vehicle?

Hönigsmann: No, the 8-MOP is usually in alcoholic solution.

Jori: Penetration is not a problem; it goes into the dermis.

Dougherty: DMSO itself penetrates.

Wilson: Perhaps the porphyrin just doesn't disaggregate.

Jori: Hp is monomeric in DMSO.

Berenbaum: meso-Tetra(hydroxyphenyl)porphyrin produces intense photosensitization of mouse dorsal skin if given systemically (Berenbaum et al 1986). Using a local application of a DMSO solution left on the skin for 45 min and then wiped off, we could produce just detectable sensitization, but we had to use high drug concentrations (20 mM) and high light doses (100 J/cm^2). Some drug penetrated to at least dermal depth, because the reaction to light included oedema and increased vascular permeability to Evans blue. However, this procedure is much less effective than systemic treatment, where a sensitizer dose of 3.125 μM and a light dose of 10 J/cm^2 produce a much greater reaction.

Kessel: If the solution was so concentrated you might not have got much light through.

Berenbaum: We drop the solution on the skin to get it to permeate. Then we wipe off any that remains on the surface before exposing the skin to light.

Jori: What is the average thickness of psoriatic plaques and how deep can light penetrate?

Greaves: They can be 0.5 cm thick.

Hönigsmann: Penetration by UVA has been measured by John Parrish (Harvard Medical School). About 20% penetrates to the lower layers of skin, which is enough to exert the therapeutic effect.

Lim: Penetration of UV is important but is not the only factor determining success of the therapy. UVB irradiation is an effective treatment for psoriasis although UVB penetrates less far than UVA: 85% of psoriasis patients respond to UVB therapy. It is likely that cytokines are involved in the therapeutic action of UVB. Irradiation of epidermal cells with UVB induces the generation of cytokines and they affect cell proliferation. Therefore, UVB doesn't have to get directly to the affected area.

Dougherty: You simply have to induce a general inflammatory reaction.

Hönigsmann: But UVB does reach the germinative layer; otherwise you could not do experiments with unscheduled DNA synthesis.

Greaves: It is increasingly thought that psoriasis is an immunological disease, from the increased numbers of T helper cells in the epidermis as well as the dermis in the early stages of development of lesions, and from the response to cyclosporin. Although we have always assumed that PUVA therapy works through the formation of DNA dimers, and that is probably its major action, I would not exclude the possibility of immunosuppressive action along the lines that Dr Ullrich has described, especially if the immunocompetent cells are present in the epidermis so that light does not have to penetrate to the dermis. There is also increased HLA-DR antigen expression in psoriasis, which is additional evidence that it may be an immunologically driven disease. This is the increasingly favoured view, rather than the older views that it is either an inflammatory disease due to eicosanoids or the result of some inherent abnormality in the epidermal cells.

Kessel: We have been working on the photosensitization of atheromatous plaques. The interesting point, which may apply to psoriasis as well, is that the porphyrins localizing in atheromatous plaques are not the same ones that localize in tumours (Kessel & Sykes 1984); the best localizer is uroporphyrin, which doesn't localize within tumours. So if there is some thought of treating psoriasis with photodynamic therapy, it may be feasible to use different porphyrins from those used to treat tumours. There are advantages in using a photosensitizing method for psoriasis that doesn't involve ultraviolet light and may not be mutagenic. The porphyrin-type photosensitizers might be useful, but we must determine which are the best dyes. Rather than the HpD-type, which persist for a long time in the skin, the short-acting sensitizers might be used.

Jori: This is why I asked about the depth of light penetration required for UVA treatment to be effective. It may not be necessary to use a porphyrin-type photosensitizer. A sensitizer absorbing at 400 nm might be just as good, but acting at the membrane level rather than on the DNA. Porphyrins are good for treating tumours because deep penetration is required, but for psoriasis we don't want too much penetration because we wish to minimize side-effects.

Dougherty: Why don't the hydrophobic portions of these drugs enter atheromatous plaques?

Kessel: The plaques are micelles—cholesterol esters with water inside—and uroporphyrin localizes in the water-rich areas. It wouldn't be possible to get into solution a dye hydrophobic enough to enter the cholesterol region.

Dougherty: You could use liposome delivery.

Kessel: We haven't tried that.

Henderson: Can you destroy atheromatous plaques with uroporphyrin?

Kessel: No, but the blood supply to the plaques is destroyed and they don't get larger.

Henderson: There is an inconsistency here. Why can't we destroy blood vessels when we inject uroporphyrin into mice? We speculate that this is because it doesn't bind to anything (Henderson & Bellnier, this volume). But you can put uroporphyrin in plaques and destroy some biological structure.

Kessel: The plaque does not get any bigger, but it is not destroyed. The only way to destroy the atheromatous plaque is with laser energy.

Henderson: What exactly does uroporphyrin destroy?

Kessel: Richard Spears (personal communication) has found evidence for a microvasculature associated with plaques, which can be destroyed by PDT.

Dougherty: In the rabbit, which is admittedly not the best model, Dr Spears saw selective destruction of the plaque with HpD and red light.

Kessel: It was the haematoporphyrin component of HpD that was being localized.

Carruth: We are interested in destroying human atheromatous plaques. We can't destroy the calcium which does not retain HpD, but we hoped to get retraction of the plaque by destroying some of the substances between the vessel wall and the calcium. The plaques can be destroyed with a laser but it is also necessary to destroy the calcium and there are problems of distal embolization and vessel wall penetration. With PDT the plaque simply retracts, which is clinically safer.

Kaye: Balloon dilatation of coronary vessels is being used, but the problem is the re-stenosis of the vessel. Therefore balloon dilatation is now being combined with an attempt to destroy the vasculature of the plaques so that they don't re-grow and re-stenose the artery.

Henderson: I still don't understand why uroporphyrin works in the one case and not in the other.

Moan: I would be surprised if uroporphyrin sensitizes a biological system to which it doesn't bind. Cells incubated in medium containing uroporphyrin do not become photosensitive, even if the concentration is extremely high. But uroporphyrin binds some biological structures, even in skin—probably connective tissue components. It may be a sensitizing effect there.

Jori: It binds to collagen, for example.

Henderson: Is there no collagen in plaques?

Kessel: A plaque is not a cell. A cell is relatively easy to kill, but a plaque requires a totally different mechanism.

Dougherty: There are cells involved, however.

Kaye: Are you saying that HpD doesn't bind in the vasculature of the plaque?

Kessel: Richard Spears sent us atheromatous plaques from rabbits that he had treated systemically with HpD. We extracted these and found that only the haematoporphyrin component of HpD binds to the plaques.

Dougherty: Perhaps your extraction was no good!

Kessel: We used the same method as we used for all our extractions. I have not tried the sulphonated tetraphenylporphyrins yet. The fluorescence signal of plaque-bound dye was consistent with a water-rich environment.

Dougherty: Perhaps you missed the important part—the part that binds to the vasculature.

Kessel: It could certainly be such a small amount that it is not detected.

Moan: Does the dye stay in the plaques for a long time? If it is in a water-rich area it should be removed quite fast. I would guess that it's not just dissolved in the water but bound to something.

Kessel: That's possible, but the fluorescence signal is certainly characteristic of a water-rich environment.

Jori: Perhaps it is bound to collagen.

Kessel: We could test that.

Dougherty: PUVA therapy has also been used in mycosis fungoides. Does that cure or control the disease?

Greaves: There is no cure for mycosis fungoides. PUVA treatment is only effective in the very early plaque stages, and not at all effective in the tumour stage, perhaps because of the penetration problems. There is no evidence that PUVA therapy alters the ultimate course of the disease.

Dougherty: Is this an immunological-based disease?

Greaves: Mycosis fungoides is thought by some to be a result of chronic unrecognized immunological stimulation which eventually leads to development of a clone of malignant T lymphocytes. The other view is that it is caused by a retrovirus.

References

Berenbaum MC, Akande SL, Bonnett R et al 1986 *meso*-Tetra(hydroxyphenyl)porphyrins, a new class of potent tumour photosensitizers with favourable selectivity. Br J Cancer 54:717–725

Black AK, Fincham N, Greaves MW, Hensby CN 1980 Time course changes in levels of arachidonic acid and prostaglandins D_2, E_2, $F_{2\alpha}$ in human skin following ultraviolet-B irradiation. Br J Clin Pharmacol 10:453–456

Chung HT, Burnham DK, Robertson B, Roberts LK, Daynes RA 1986 Involvement of prostaglandins in the immune alterations caused by exposure of mice to ultraviolet radiation. J Immunol 137:2478–2484

DeLeo VA, Hanson D, Weinstein IB, Harber LC 1985 Ultraviolet radiation stimulates the release of arachidonic acid from mammalian cells in culture. Photochem Photobiol 41:51–56

De Mol NJ, Beijersbergen van Henegouwen GMJ, Mohn GR, Glickman BW, Van Kleef PM 1981 On the involvement of singlet oxygen in mutation induction by 8-methoxypsoralen and UVA radiation in *Escherichia coli* K-12. Mutation Res 82:23–27

Diezel W, Meffert H, Sönnichsen N 1980 Therapy of psoriasis with hematoporphyrin derivative and long wave ultraviolet light. Dermatol Monatsschr 166:793–797

Gold RL, Anderson RR, Natoli VD, Gange RW 1988 An action spectrum for photoinduction of prolonged cutaneous photosensitivity by topical 8-MOP. J Invest Dermatol 90:818–822

Gruner S, Meffert H, Volk H-D, Grunow R, Jahn S 1985 The influence of haematoporphyrin derivative and visible light on murine skin graft survival, epidermal Langerhans cells and stimulation of the allogeneic mixed leucocyte reaction. Scand J Immunol 21:267–273

Henderson BW, Bellnier DA 1989 Tissue localization of photosensitizers and the mechanism of photodynamic tissue destruction. In: Photosensitizing compounds: their chemistry, biology and clinical use. Wiley, Chichester (Ciba Found Symp 146) p 112–130

Kessel D, Sykes E 1984 Porphyrin accumulation by atheromatous plaques of the aorta. Photochem Photobiol 40:59–61

LeVee GJ, Applegate LA, Ley RD 1988 Photoreversal of the UV radiation-induced disappearance of ATPase positive Langerhans cells in the epidermis of *Monodelphis domestica*. J Leukocyte Biol 44:508–513

Plummer NA, Hensby CN, Warin AP, Camp RD, Greaves MW 1978 Prostaglandins E_2, $F_{2\alpha}$ and arachidonic acid levels in irradiated and unirradiated skin of psoriatic patients receiving PUVA treatment. Clin Exp Dermatol 3:367–369

Robertson B, Gahring L, Newton R, Daynes RA 1987 *In vivo* administration of IL-1 to normal mice decreases their capacity to elicit contact hypersensitivity responses: prostaglandins are involved in this modification of the immune response. J Invest Dermatol 88:380–387

Rubin DB 1988 Prostaglandins and radiation in cell culture. In: Polgar P (ed) Eicosanoids and radiation. Kluwer Academic Publishers, Boston, p 13–24

Schwarz T, Urbanska A, Gschnait F, Luger TA 1986 Inhibition of the induction of contact hypersensitivity by a UV-mediated epidermal cytokine. J Invest Dermatol 87:289–291

Effects of porphyrins on skin

Henry W. Lim

Dermatology Service, New York Veterans Administration Medical Center, Department of Dermatology, New York University School of Medicine, 408 First Ave., New York, NY 10010, USA

Abstract. Porphyrins and radiation induce activation of the complement system, release of mediators from mast cells, dermal accumulation of neutrophils, and activation of factor XII-dependent pathways. Recent studies have also shown that porphyrins and radiation have profound effects on the eicosanoid metabolism of mast cells, endothelial cells, macrophages and epidermal and dermal homogenates. *In vitro*, prostaglandin D_2 (PGD_2) was generated when mast cells were exposed to protoporphyrin and radiation. Exposure of cultured human umbilical vein endothelial cells or human omental microvascular endothelial cells to uroporphyrin or protoporphyrin and radiation resulted in predominant generation of $PGF_{2\alpha}$, while PGE was generated after the exposure of macrophages to Photofrin II and radiation. Activities of epidermal eicosanoid-metabolizing enzymes in protoporphyric mice were suppressed following *in vivo* exposure to radiation, whereas activities of the dermal enzymes were enhanced 18 h after radiation. These effects of porphyrins on skin contribute to the development of porphyrin-induced cutaneous phototoxicity, which is seen in patients with porphyrias and as a consequence of photodynamic therapy.

1989 Photosensitizing Compounds: their Chemistry, Biology and Clinical Use. Wiley, Chichester (Ciba Foundation Symposium 146) p 148–158

Cutaneous phototoxicity, consisting of erythema, oedema, vesicles and erosions, occurs as a consequence of exposure to porphyrins and radiation. These changes can be seen in patients with porphyrias and in those receiving photodynamic therapy. To further elucidate the pathophysiology of porphyrin-induced phototoxicity, the interactions of porphyrins, radiation, and the various constituents of the skin have been studied. The results are reviewed and summarized here.

Porphyrins and the complement system

The complement system consists of a group of serum proteins which participates actively in inflammation. Complement activation results in membrane damage and cytolysis through the formation of the membrane attack complex (C5b–9), as well as the generation of biologically active products (anaphylatoxins C3a, C4a and C5a).

Studies on the interactions of porphyrins and the complement system have been performed in patients with erythropoietic protoporphyria (EPP), where protoporphyrin is the predominant porphyrin, and in patients with porphyria cutanea tarda (PCT), where uroporphyrin is elevated. In patients with these porphyrias, deposition of complement components at the vessel walls and dermoepidermal junction has been reported (Cormane et al 1971, Epstein et al 1973). After irradiation of the skin of patients with EPP, in addition to endothelial cell damage, degranulation of mast cells and infiltration of polymorphonuclear cells within and around dermal blood vessels were observed (Schnait et al 1975). These are histological changes consistent with those mediated by complement activation products. The participation of complement in porphyrias was corroborated by the following studies. Complement activation, as assessed by measurement of haemolytic activities and generation of complement-derived chemotactic activity, was demonstrated in sera obtained from patients with EPP and PCT exposed to radiation *in vitro* and *in vivo* (Lim et al 1981, 1984). In animal models, the development of porphyrin-induced phototoxicity was associated with a decrease in the total complement haemolytic activity (CH50), and the phototoxic response was suppressed in complement-depleted animals as well as in animals congenitally deficient in C5 (Kamide et al 1984, Lim & Gigli 1981, Lim et al 1985, 1986a).

Porphyrins and mast cells

Mast cells are present in perivascular and periappendageal locations in the dermis. Upon activation, they release vasoactive mediators, chemotactic factors and various enzymes; therefore they play an important role in cutaneous inflammation (Soter & Austen 1987).

As mentioned, mast cell degranulation was noted in the exposed skin of patients with EPP (Schnait et al 1975). *In vitro* studies further corroborated the role of mast cells in porphyrin-induced phototoxicity. Exposure of purified rat peritoneal or mouse bone marrow-derived mast cells to clinically relevant doses of protoporphyrin and irradiation *in vitro* resulted in the release of preformed and generated mediators. This release was independent of calcium, associated with membrane perturbation, and mediated in part by hydrogen peroxide, which is known to be generated by 'excited state' porphyrins (Lim et al 1986b, 1987, Glover et al 1988). In contrast, no mediator release was observed in the presence of uroporphyrin and radiation (Lim et al 1987). These findings may partly explain the differences in the clinical presentation of patients with EPP and PCT. The elevated protoporphyrin in patients with EPP may induce the release of mast cell mediators *in vivo* after exposure to the sun, resulting in erythema, oedema, urticarial lesions and pruritus. The absence of these changes in patients with PCT may be explained by the lack of direct effect of uroporphyrin on cutaneous mast cells.

The participation of mast cells in porphyrin-induced phototoxicity was further confirmed in studies performed in animal models. In mice, haematoporphyrin derivative-induced phototoxicity was associated with biphasic elevation of serum histamine levels (at 1 h and at 12 h after irradiation), dermal mast cell degranulation (which peaked at 12 h) and an increase in the number of hypogranulated mast cells at 18–48 h. These changes were associated with a marked dermal neutrophilic infiltrate (Kerdel et al 1987, He et al 1989). In animal models, the porphyrin-induced phototoxic response was suppressed by pretreatment with antihistamines, and by intradermal injection of a mast cell secretagogue compound 48/80 (Kamide et al 1984, Lim et al 1985). Furthermore, mice congenitally deficient in mast cells had a much more diminished phototoxic response than mast cell-sufficient animals (Lim et al 1986a).

Porphyrins and polymorphonuclear leucocytes

Polymorphonuclear leucocytes (PML) are one of the major effector cell types in inflammation. They are recruited to the sites of inflammation by neutrophil chemotactic factors, which can be generated as a result of complement activation, degranulation of mast cells, and metabolism of eicosanoids. Activated PML modulate the local inflammatory response by the release of enzymes and generation of eicosanoids and reactive oxygen species (Goetzl & Goldstein 1989).

The possible role of PML in phototoxicity was suggested by the observation in animal models that porphyrin-induced phototoxic response was associated with dermal PML infiltrate (He et al 1989, Baart De La Faille et al 1980). The fact that the phototoxic response in leukopenic animals was markedly suppressed compared to that in control animals further indicated that PML were required for the complete manifestation of phototoxicity induced by porphyrins (Lim et al 1985, 1986a). *In vitro* exposure of human PML to protoporphyrin and radiation resulted in membrane damage; in contrast, uroporphyrin and radiation did not induce any alterations (Sandberg et al 1981), results which were similar to those obtained in studies with mast cells.

Porphyrins and factor XII-dependent pathways of coagulation

Activation of factor XII (Hageman factor)-dependent pathways of coagulation results in the generation of kallikrein, which is chemotactic for neutrophils. Kallikrein cleaves high molecular weight kininogen to a proinflammatory peptide, bradykinin. The properties of bradykinin include smooth muscle contraction, vasodilatation, and production of pain upon contact with sensory nerve endings (Kaplan 1985). Therefore, activation of these pathways can contribute to cutaneous inflammation.

In an *in vitro* study, it was shown that protoporphyrin induced activation of Hageman factor-dependent pathways with the generation of kallikrein activity

(Becker et al 1985). In contrast, neither uroporphyrin nor coproporphyrin induced such activation. Whether this activation, which was independent of irradiation, contributes to the pathogenesis of porphyrin-induced phototoxicity remains to be further investigated.

Porphyrins and the eicosanoids

The eicosanoids are biologically active, unstored, oxygenated metabolites of the twenty-carbon essential fatty acid arachidonic acid (Waldman & Lim 1988). They are synthesized in cells in response to physical or biochemical stimuli. Once the precursor arachidonic acid is released from the membranes, it is rapidly metabolized by one of two pathways: cyclooxygenase, resulting in the generation of the prostaglandins (PGs) and thromboxanes; or lipoxygenase, resulting in the formation of leukotrienes and hydroxy acids. The effects of eicosanoids in the skin include regulation of vascular tone, permeability of capillaries and venules, epidermal proliferation and leucocyte chemotaxis.

The effects of porphyrins and radiation on eicosanoid metabolism have been examined in several studies. Exposure of purified rat mast cells to protoporphyrin and radiation resulted in the generation of prostaglandin D_2 (PGD_2), with lesser amounts of hydroxy acids, PGE_2 and 6-keto-$PGF_{1\alpha}$ (a stable metabolite of PGI_2) (Lim et al 1986b). Incubation of mouse peritoneal macrophages or RIF tumour cells with Photofrin II followed by 630 nm radiation resulted in dose-dependent generation of PGE (Henderson & Donovan 1988).

The effects of uroporphyrin (Uro, 250–5000 ng/ml) and protoporphyrin (Pp, 250–1000 ng/ml) on human umbilical vein endothelial cells and human omental microvascular endothelial cells have also been examined (Lim & Sarmalkar 1989). Exposure of ^{3}H-arachidonic acid-labelled endothelial cells to Uro resulted in Uro-dependent net releases of 7–19% of the radioactivity immediately after the completion of irradiation, and net releases of 21–38% an hour later. No correlation between the releases and cell viability was noted, suggesting that the releases were not due to cell death. Similar findings were noted with Pp. Catalase markedly suppressed the release of radioactivity, whereas mannitol, sodium azide, and superoxide dismutase did not, indicating the importance of hydrogen peroxide in this process. Radio-thin-layer chromatographic analysis of the eicosanoid profile revealed that while the calcium ionophore A23187 induced predominant release of 6-keto-$PGF_{1\alpha}$ and $PGF_{2\alpha}$, Uro or Pp and radiation induced a selective release of $PGF_{2\alpha}$.

To investigate the activities of cutaneous eicosanoid metabolizing enzymes in porphyria, we rendered Skh: HR-1 hairless albino mice protoporphyric by feeding them 0.5% collidine (He & Lim 1989). Phototoxicity was induced by exposing the mice to 12 kJ/m^2 of 396–406 nm radiation. At 0, 6, 12, and 18 h after radiation, the skin between the scapulae and iliac crests was separated into epidermis and dermis and then homogenized. The activities of the cyclooxygenase and the lipoxygenases in homogenates were determined by incubation with ^{3}H-arachidonic acid, and the eicosanoids generated were characterized by radio-thin-layer chromatography. Epidermis of unirradiated normal mice generated 6-keto-$PGF_{1\alpha}$, $PGF_{2\alpha}$, PGE_2, PGD_2 and hydroxy-eicosatetraenoates (HETEs) at 18 h; similar values were observed in irradiated normal mice and in unirradiated porphyric mice. In contrast, generation of eicosanoids by the epidermis of irradiated porphyric mice was markedly suppressed at all the time-points examined. The dermis of irradiated porphyric mice generated increased amounts of PGE_2 (7.2%, vs control of 1.8%) and HETEs (33.2%, vs control of 7.7%) at 18 h, which is probably due to the polymorphonuclear cell infiltrate observed in the dermis at this time. These results demonstrate that irradiation of protoporphyric mice resulted in alteration of the activities of cutaneous eicosanoid metabolizing enzymes, which may be of pathophysiological importance in the development of phototoxicity in protoporphyric mice.

Conclusion

From the studies reviewed, it is clear porphyrins and radiation have profound effects on skin which contribute to the pathogenesis of cutaneous changes observed in porphyrin-induced phototoxicity. Exposure of porphyrins to the appropriate radiation results in the generation of biologically active complement activation products, which can induce erythema, oedema, mast cell degranulation, and polymorphonuclear cell infiltration. The direct effect of porphyrins and radiation on the release of mast cell-derived mediators, and on the alteration of eicosanoid metabolism, may also contribute to the development of porphyrin-induced phototoxic lesions.

Acknowledgement

This work was supported by research grants from the National Institutes of Health (AR 36624) and the Veterans Administration.

References

Baart De La Faille H, Woutersen RA, Van Weelden H, Baart De La Faille-Kuyper EH 1980 Experimental protoporphyria in hairless mice: photodermatitis. In: Willoughby DA, Giroud JP (eds) Inflammation: mechanisms and treatment. University Park Press, Baltimore, p 603–615

Becker CG, Wagner M, Kaplan A et al 1985 Activation of factor XII-dependent pathways in human plasma by hematin and protoporphyrin. J Clin Invest 76:413–419

Cormane RH, Szabo E, Hoo TT 1971 Histopathology of the skin in acquired and hereditary porphyria cutanea tarda. Br J Dermatol 85:531–539

Epstein JH, Tuffanelli DH, Epstein WL 1973 Cutaneous changes in the porphyrias. Arch Dermatol 107:689–698
Glover RA, Barrett KE, Wasserman SI et al 1988 A possible role for hydrogen peroxide in phototoxic histamine release from mast cells induced by protoporphyrin and UVA light. Clin Res 36:469A
Goetzl EJ, Goldstein IM 1989 Granulocytes. In: Kelley WN et al (eds) Textbook of rheumatology. Third edition. WB Saunders Co, Philadelphia, p 322–345
He D, Lim HW 1989 Alteration of cutaneous eicosanoid metabolism in protoporphyric mice. Clin Res 37:698A
He D, Soter NA, Lim HW 1989 The late phase of hematoporphyrin derivative-induced phototoxicity in mice: release of histamine and histologic changes. Photochem Photobiol 50:91–95
Henderson BW, Donovan JM 1988 Prostaglandin release from peritoneal macrophages and tumor cells after photodynamic therapy. Photochem Photobiol 47:76s
Kamide R, Gigli I, Lim HW 1984 Participation of mast cells and complement in the immediate phase of hematoporphyrin-induced phototoxicity. J Invest Dermatol 82:485–490
Kaplan AP 1985 The intrinsic coagulation, fibrinolytic, and kinin-forming pathways of man. In: Kelly WN et al (eds) Textbook of rheumatology. Second Edition. WB Saunders Co, Philadelphia, p 95–115
Kerdel F, Soter NA, Lim HW 1987 In vivo mediator release and degranulation of mast cells in hematoporphyrin derivative-induced phototoxicity. J Invest Dermatol 88:277–280
Lim HW, Gigli I 1981 Role of complement in porphyrin-induced photosensitivity. J Invest Dermatol 76:4–9
Lim HW, Sarmalkar M 1989 Effect of porphyrins on endothelial cells eicosanoid metabolism. Clin Res 37:700A
Lim HW, Perez HD, Poh-Fitzpatrick M, Goldstein IM, Gigli I 1981 Generation of chemotactic activity in serum from patients with erythropoietic protoporphyria and porphyria cutanea tarda. N Engl J Med 304:212–216
Lim HW, Poh-Fitzpatrick MB, Gigli I 1984 Activation of the complement system and generation of chemotactic activity in vivo in patients with porphyrias. J Clin Invest 74:1961–1965
Lim HW, Young L, Hagan M, Gigli I 1985 Delayed phase of hematoporphyrin-induced photoxicity: modulation by complement, leukocytes and antihistamines. J Invest Dermatol 84:114–117
Lim HW, Hagan M, Gigli I 1986a Phototoxicity induced by hematoporphyrin derivative in C5-deficient, mast cell-deficient and leukopenic mice. Photochem Photobiol 44:175–180
Lim HW, Parker D, Marcus AJ 1986b Generation of eicosanoids from mast cells exposed to protoporphyrin and irradiation. Clin Res 34:763A
Lim HW, Wasserman SI, Gigli I 1987 Differential effects of protoporphyrin and uroporphyrin on murine mast cells. J Invest Dermatol 88:281–286
Sandberg S, Glette J, Hopen J, Solberg CO, Romslo I 1981 Porphyrin-induced photodamage to isolated human neutrophils. Photochem Photobiol 34:471–475
Schnait FG, Wolff K, Konrad K 1975 Erythropoietic protoporphyria—submicroscopic events during the acute photosensitivity flare. Br J Dermatol 92:545–557
Soter NA, Austen KF 1987 Mast cells. In: Fitzpatrick TB et al (eds) Dermatology in general medicine. McGraw-Hill, New York, p 426–435
Waldman JS, Lim HW 1988 Effects of ultraviolet radiation on eicosanoid metabolism in intact skin. In: Polgar P (ed) Eicosanoids and radiation. Academic Publishers, Boston, p 61–77

DISCUSSION

Dougherty: Why does uroporphyrin behave differently when intravenously injected and in endothelial cells? You observed some cell death in the latter system.

Lim: It probably depends on concentration. I was surprised by the results in endothelial cells, because in mast cells uroporphyrin was completely inert.

Dougherty: You used a blue light source and therefore the Soret band for activation. Most of us use red light. Using the Soret band means using a higher dose of light.

Moan: I don't think that is the explanation. It must have something to do with the types of cell. Is serum present in the incubation medium?

Lim: We have to label the cells with arachidonic acid. Therefore we use a serum-free medium.

Moan: How long are the cells in a serum-free medium?

Lim: We use an initial incubation of about 4 h and then we wash the cells and incubate them for another 30 min.

Moan: Usually cells suffer markedly from such a long time without serum.

Dougherty: Could the cells therefore be partially damaged initially?

Henderson: We tried various conditions for our studies on prostaglandin E (PGE) in RIF tumour cells (unpublished work). We kept the cells *in vitro* in 1% serum, in medium with no serum, or in Hanks for up to three hours. These conditions alone did not induce prostaglandin release.

Lim: The question is whether the serum-free conditions cause the cells to be fragile.

Henderson: If the cell membrane were fragile some prostaglandin release would be triggered in that time. The only difference that I find from your studies was that prostaglandin release was completely correlated with cell death, based on Trypan blue exclusion, which is a good way for looking for membrane damage with PDT.

Jori: The Trypan blue test can give variable results depending on how you use it. When do you assay cells for Trypan blue exclusion?

Lim: Immediately at the end of the irradiation.

Henderson: That may account for the discrepancy. I did it in a time-dependent manner. The increase in prostaglandin release exactly coincides with the increase in Trypan blue uptake. Twenty minutes after the end of PDT we see maximum Trypan blue uptake and maximum PGE release.

Greaves: You observed raised levels of histamine reaching a maximum 12 h after irradiation of HpD-treated mice. In your *in vitro* studies with isolated mast cells what incubation time produced the release of preformed mediators?

Lim: We labelled the mast cells with radiolabelled serotonin and detected the release of labelled serotonin, as an indicator of histamine release, immediately after completion of 20 min of irradiation.

Greaves: Why did it take 12 h for maximum release *in vivo*?

Lim: There was an earlier peak, after about 60 min. That decreased and then a second peak appeared, reaching a maximum at 12 h. One possible explanation is that during the development of phototoxicity lymphocytes and polymorphonuclear cells infiltrate and generate histamine-producing factors. There might be an interaction between the infiltrating cells and the mast cells that are already present, resulting in the eventual release of histamine.

Henderson: It has been suggested that polymorphs and prostaglandins are necessary for oedema formation.

Lim: Wedmore & Williams (1981) found that inflammatory mediators such as $C5_a$, leukotriene B_4, and bradykinin increased vascular permeability, and the presence of PGE_2 markedly enhanced this effect. The increase in vascular permeability was markedly suppressed when the animals were rendered neutropenic. Therefore, there is a synergistic effect between polymorphonuclear cells, prostaglandins and other inflammatory mediators. In leukopenic animals there was a marked suppression of the phototoxic response.

Dougherty: How can we protect patients from cutaneous sensitivity?

Lim: Antihistamines don't work very well. Sunscreens, as we discussed earlier (p 16), protect against the wrong spectrum. We don't have an answer yet. The problem is that many factors interact to cause cutaneous phototoxicity. Steroids can modulate and down-regulate most of these mediators but they have significant side-effects and are not appropriate for use in a recurrent or chronic fashion.

Dougherty: How about giving antihistamine and aspirin? Is that innocuous enough to try?

Lim: Yes, and that may be useful. One could try an aspirin every four hours for 24 h before the therapy. I don't have any data on this.

Greaves: You can block early phase UVB erythema in man with therapeutic doses of indomethacin. I am not sure about aspirin. Apart from the effects of putting nicotinic acid ester on the skin, which also produces inflammation that is blockable by non-steroidal compounds, I don't know of any other form of inflammation that's affected by non-steroidal anti-inflammatory drugs or whether anyone has tried them on inflammation caused by porphyrin photosensitization.

Lim: Indomethacin is also effective when applied locally; it absorbs UVA and acts as a sunscreen. I don't know about the long-term efficacy or safety of its topical application.

Dougherty: What kind of vehicle is used for it?

Lim: Indomethacin is dissolved in propylene glycol: absolute ethanol: dimethylacetamide (19:19:2) (Lim et al 1983).

Henderson: We thought indomethacin might be beneficial in inhibiting the vascular effects of PDT. By treating mice with indomethacin (5 mg/kg daily for four days) followed by PDT (25 mg/kg Photofrin II, 135 J/cm^2, 630 nm), we could prevent vascular shutdown in tumour and skin. However, we noticed that indomethacin treatment reduced tumour angiogenesis, i.e. the tumours had a reduced blood supply. We used [^{14}C]Photofrin II to study uptake in skin and tumours and found that the uptake was halved in indomethacin-treated mice. Thus the decreased vascular PDT effects in these animals may have been caused simply by decreased sensitizer uptake.

Lim: What happens if you give the indomethacin very late?

Henderson: Then we observed only minimal effects. However, T. J. Wieman (personal communication) reports inhibition of vascular PDT effects under these conditions in rats. But he didn't measure porphyrin levels.

Jori: Why should indomethacin inhibit porphyrin uptake?

Henderson: I have no idea.

Dougherty: It also did that in the liver.

Greaves: Indomethacin doesn't just affect cyclooxygenase; it stabilizes cell membranes. Particularly at high doses, this could be relevant in the inhibition of porphyrin uptake.

Dougherty: Would it inhibit endocytosis?

Greaves: I'm not sure what the biological consequences of stabilization of the cell membranes would be.

Jori: Did you measure the serum levels of porphyrins?

Henderson: No.

Carruth: How about removing the porphyrin more rapidly? I have heard unconfirmed reports that diuretics can reduce skin sensitivity (B. Aronoff, personal communication).

Lim: Charcoal is also used in an attempt to clear the drug quickly.

Dougherty: We tested charcoal in mice. It had a small effect, but it wasn't worth pursuing. The assumption was that the charcoal interferes with recirculation of the drug in the liver.

Lim: That is used clinically in some patients with porphyria. A large amount of charcoal is required to interrupt the enterohepatic circulation of the porphyrins. The drugs are absorbed out and then excreted.

Brown: We are trying that with PDT patients. It's too early to say what the results are, but they are not dramatically good.

Moan: What about escalating the light doses? A low dose given one day may lead to development of protective mechanisms, such as skin thickening.

Dougherty: No, that's not what happens. In our animal studies we purposely picked initial doses that didn't cause any measurable therapeutic effect. The next day we could double the dose without inducing a therapeutic effect because part of the drug had gone. This is, we believe, a photobleaching phenomenon.

Moan: That's not necessarily a bleaching effect. It might be related to oedema or skin thickening.

Dougherty: We reinjected the animal with the same dose and the effect on the original foot was exactly as predicted for no previous treatment.

Berenbaum: Did you compare the effect of a second dose of light given immediately after the first, rather than 24 h later?

Dougherty: No, we didn't try that.

Moan: That's important because the bleaching effect would be similar in both cases.

Dougherty: You can demonstrate photobleaching by looking at the fluorescence emission.

Berenbaum: Photobleaching may occur, but may not explain your observations. If the effect is caused by photobleaching, immediate repetition of the light dose will produce the same response as would be produced 24 h later.

Dougherty: That's a good suggestion. But I think it has to be photobleaching. We can demonstrate a reduction in the amount of drug in the tissue, either by fluorescence or by extraction.

Truscott: Couldn't it be migrating away during irradiation?

Moan: Yes; the low light dose that Dr Dougherty used may not bleach the dye.

Dougherty: We predict that 50–65% of the drug is bleached with that single dose.

Moan: That will not happen *in vitro*.

Dougherty: No, that is correct.

Moan: The explanation may be, as Dr Truscott says, a release of the drug. We have shown that exposure to light of porphyrins bound to cells results in a destruction of the binding sites followed by redistribution of the porphyrin molecules (Moan et al 1988).

Berenbaum: That doesn't matter.

Dougherty: It isn't practical to use this dose escalation clinically because most of our patients are not local; we would have to keep them in the hospital.

Brown: Light dose escalation might happen accidentally over a long period. That could make a significant contribution to the disappearance of the drug in the long term.

Dougherty: We have long suspected that patients who do not exclusively stay out of the sun probably become desensitized more quickly, but obviously we can't control that.

Wilson: Has anyone compared mice kept completely in the dark for a long period with controls kept under normal ambient lighting?

Dougherty: We did that. The results appeared to be consistent with what we suspected, but it wasn't a greatly controlled experiment.

Hönigsmann: The use of chloroquine is a different issue. Nobody knows how it works in porphyria cutanea tarda (PCT).

Lim: Chloroquine is thought to help remove porphyrins from the liver to be excreted into the urine. That is why one has to be careful when using chloroquine; the patient can go into crisis because the amount of porphyrin in serum can increase tremendously.

Hönigsmann: But that doesn't explain why people can stay free of PCT for

2–3 years after one course of chloroquine over 2–3 months. They are free of disease but do not excrete increased amounts of uroporphyrin. The mechanism is unknown.

Greaves: Don't you deplete the hepatic stores of porphyrin when you use chloroquine? Normally you would venesect the patient first in an attempt to deplete the stores. Patients given chloroquine don't suffer the normal acute exacerbation of the disease.

Jori: How could one demonstrate a correlation between porphyrin levels in the serum and skin photosensitivity?

Brown: One can try to correlate the results of skin patch testing and measurements of serum levels by standard fluorescence.

Lim: It is difficult to induce phototoxic responses in the skin of porphyric patients with artificial light. One has to use sunlight, which complicates the method.

Dougherty: We don't find any enhanced skin reaction when we treat patients with PDT at times when the serum levels of the injected drug are highest.

Brown: There's not a clear-cut correlation; one must use statistics. That means studying many patients. I can't tell you the answer yet.

References

Lim HW, Thorbecke, GJ, Baer RL, Gigli I 1983 Effect of indomethacin on alteration of ATPase-positive Langerhans cell density and cutaneous sunburn reaction induced by ultraviolet-B radiation. J Invest Dermatol 81:455–458

Moan J, Rimington C, Malik Z 1988 Photoinduced degradation and modification of Photofrin II in cells in vitro. Photochem Photobiol 47:363–367

Wedmore CV, Williams TJ 1981 Control of vascular permeability by polymorphonuclear leukocytes in inflammation. Nature (Lond) 289:646–650

Photosensitizing compounds in the treatment of psoriasis

Herbert Hönigsmann, Adrian Tanew, Johannes Brücke and Berhard Ortel

Division of Photobiology, Department of Dermatology I, University of Vienna, Alser Strasse 4, A1090 Vienna, Austria

Abstract. Photosensitizers were first used to treat psoriasis 15 years ago when the phototoxic reaction of psoralens and UVA was found to induce remissions of the disease. The effect of this reaction on DNA, particularly the formation of cross-links, was thought to be the decisive event. Strong cross-linking agents such as 8-MOP, TMP and 5-MOP are clinically effective whereas most compounds which produce only monofunctional adducts are virtually ineffective. Orally administered 8-methoxypsoralen (8-MOP) is the most widely used compound. 4,5′,8-Trimethylpsoralen (TMP) is poorly absorbed from the intestine but has marked efficacy when applied topically. 5-MOP may be a useful alternative to 8-MOP because it is less erythemogenic and does not cause nausea. These three furocoumarins appear to be similar photochemically and may introduce similar risks. However, the photobiological properties of furocoumarins can be modified by altering one or more parts of the molecule. Such modifications might yield effective analogues with reduced cytogenetic hazards. Several psoralens and angular furocoumarins are being tested for effectiveness combined with fewest long-term side-effects, especially carcinogenesis. Encouraging preliminary results have been obtained with 7-methylpyridopsoralen and 4,6,4′-trimethylangelicin. Other important approaches to increasing the safety of photochemotherapy may be the use of different photoactivating wavelengths or the introduction of new classes of photosensitizers.

1989 Photosensitizing Compounds: their Chemistry, Biology and Clinical Use. Wiley, Chichester (Ciba Foundation Symposium 146) p 159–170

Psoriasis is a genetically determined skin disease of unknown pathogenesis. It can be a severely physically and socially disabling condition. A variety of treatments can induce complete clearing of skin lesions but the disease usually recurs after a variable disease-free interval and some patients require long-term maintenance therapy to stay in remission for a reasonable period of time. All treatments currently in use are potentially cumulatively toxic. Therefore the present policy should be to cycle patients through different treatments, alone or in combination, in order to remain below the critical toxicity of each treatment. Photochemotherapy regimens, with or without adjunctive treatments, are the most convenient and acceptable treatments for the patients themselves.

Photosensitizers were introduced into the treatment of psoriasis in 1974 when, on the basis of previous experimental work, the combination of some furocoumarins (psoralens) and UVA irradiation was found to induce remissions of the disease after repeated, controlled phototoxic skin reactions (Parrish et al 1974). These reactions occur when psoralens are photoactivated by UVA. Psoralens or UVA alone are ineffective. Because of the penetration characteristics of UVA, photochemistry is confined to the skin and thus psoralen photochemotherapy (PUVA) can be considered to be targeted.

The mechanisms of the therapeutic effect on psoriasis are unknown. Most hypotheses have stressed that UVA causes the conjugation of psoralens to DNA and the subsequent suppression of DNA synthesis and cell division (Epstein & Fukuyama 1975, Fritsch et al 1979, Walter et al 1973). It is generally assumed, although not proven, that the formation of psoralen–DNA interstrand cross-links is the decisive event responsible for suppressing the epidermal hyperproliferative stimulus in psoriasis (Parrish et al 1974, Wolff et al 1976). However, the successful use of PUVA in non-hyperproliferative skin diseases is unlikely to be attributable to this molecular reaction (Hönigsmann et al 1987). It is possible that PUVA may affect other cells in the skin, such as polymorphonuclear leucocytes or lymphocytes, and there is reason to speculate that there may be an action on immunocompetent cells in the cutaneous infiltrate. Yet, until recently, clinical experience has indicated that only cross-linking psoralens are effective in psoriasis (Hönigsmann et al 1987). Therefore only cross-linking linear furocoumarins, namely 8-methoxypsoralen (8-MOP) and 4,5′,8-trimethylpsoralen (TMP), have become established as standard photosensitizers in photochemotherapy. 8-MOP is the most widely used compound and is mainly administered orally for systemic treatment. TMP is used topically, because of its poor intestinal absorption, in the form of PUVA baths for whole-body treatment (Fischer & Alsins 1976, Turjanmaa et al 1985).

Both forms of treatment are very effective in the management of psoriasis. However, there are restrictions to their use. 8-MOP is often poorly tolerated by patients, causing nausea and vomiting. The practicability and acceptability of the TMP bath treatment, on the other hand, is hampered in many centres by the logistics of the bathing procedure. With both forms of treatment dosimetry may become a problem—particularly when one is dealing with light-sensitive patients—and overdosage may occur. This manifests itself as delayed sunburn-like erythema and skin inflammation which may progress to blistering and superficial necrosis.

Besides acute side-effects, there are always concerns about long-term risks, particularly carcinogenesis (Henseler et al 1987, Stern et al 1984, Stern & Lange 1988). Consequently, new psoralens and angular furocoumarins have been developed (Dubertret et al 1985, Guiotto et al 1984) and tested with the aim of expanding the therapeutic range of the photosensitizer, and of reducing short-term side-effects and/or long-term hazards (Tanew et al 1988b).

Methods of treatment

The general principle is to keep the concentration of the substance or the oral dose of the drug constant and to vary the UVA irradiation according to the individual sensitivity of the patient. Patients are usually subjected to whole-body irradiation by an appropriate UVA system with an emission maximum at 365 nm. The interval between application of the drug and irradiation depends on the pharmacokinetics of the substance (Hönigsmann et al 1982, Tanew et al 1988a); treatments are given when the concentration reaches its maximum in the skin.

The initial UVA doses are predetermined by phototoxicity testing before therapy (Wolff et al 1977). Repeated exposures (three to four times per week) are required to clear psoriasis and, as pigmentation develops, irradiation doses have to be adjusted to maintain a constant therapeutic effect. The use of low doses often results in treatment failures because the increasing pigmentation acts as a natural protective screen against phototoxicity (Hönigsmann et al 1987). It has become apparent from several large-scale studies that an aggressive but personalized treatment with appropriate dose adjustments saves time and reduces the total cumulative UVA energy load as compared with a cautious but rigid dosimetry scheme (Wolff & Hönigsmann 1981).

Treatment results

The efficacy of photochemotherapy with oral 8-MOP in inducing and maintaining remissions of psoriasis has been widely documented and confirmed by multicentre trials in Europe and the United States involving over 5000 patients (Henseler et al 1981, Clinical Cooperative Study 1979, Melski et al 1977). Excellent results with clearing rates of 85 to 90% after an average of 20 to 25 exposures have been reported. Similarly satisfying results have been presented by Scandinavian groups with TMP baths and total-body UVA irradiation (Turjanmaa et al 1985).

When the psoriasis is completely cleared, patients are assigned to maintenance therapy (Wolff et al 1976), with reduced treatment frequency, which is essential to prevent early recurrences. Follow-up observations have indicated that the majority of patients who receive maintenance therapy for about two months remain free of disease for at least six to 12 months (Wolff & Hönigsmann 1981).

Side-effects

Acute short-term side-effects

These effects are due to either overdosage of UVA or 8-MOP intolerance (Parrish et al 1974, Wolff et al 1976). Most frequently, overexposure results in phototoxic erythema which may finally lead to oedema and blistering. Pruritus is clearly dose-related and may sometimes be manifest as an intense, torturing itch, particularly in severely overdosed skin (Wolff & Hönigsmann 1981). Transient nausea occurs in more than 10% of patients after oral administration of 8-MOP. If nausea is accompanied by vomiting, the symptoms may be severe enough to necessitate discontinuation of photochemotherapy.

Chronic long-term side-effects

While acute effects are reversible after discontinuation of photochemotherapy, long-term side-effects represent an area of greater concern. From *in vitro* studies and animal experiments with 8-MOP it appears that prolonged repeated photochemotherapy may result in ocular damage, immunological alterations, chronic actinic skin damage and cutaneous carcinogenesis. It seems reasonable to assume that these potential risks depend on the photochemical properties of the photosensitizer and on the total UVA dose.

The risk of greatest concern is that of initiating or promoting skin cancers. The photochemical reaction of certain psoralens, particularly cross-linking analogues such as 8-MOP, has long been known to produce mutations in bacteria (Igali et al 1970, Swanbeck & Thyresson 1974) and in mammalian cells (Schenley & Hsie 1981). The phototoxic reaction of 8-MOP is carcinogenic in animals under certain experimental conditions and there is evidence that some groups of patients have an elevated risk of non-melanoma skin cancer after prolonged PUVA treatment. This issue has been extensively reviewed by Hönigsmann et al (1987) and Stern & Lange (1988).

Improvements in photochemotherapy

Improvements in photochemotherapy must aim at a reduction of both short- and long-term side-effects. New approaches include the development of treatment schedules which permit reduction in the frequency of treatments and in the total cumulative UVA dose, the introduction of new compounds with reduced cytogenetic hazards, and the combination of photochemotherapy with topical and/or systemic adjuvant therapy.

We shall summarize here our preliminary clinical results with three furocoumarins, 5-methoxypsoralen (5-MOP), 7-methylpyridopsoralen (7-MPP) and 4,6,4′-trimethylangelicin (4,6,4′-TMA). In addition, we shall discuss possible improvement in therapy by using wavelengths other than 365 nm.

5-Methoxypsoralen

5-MOP is a naturally occurring linear psoralen that is known to be clinically effective in psoriasis. From previous studies it appeared to be less phototoxic than 8-MOP and thus safer with regard to short-term side-effects (Hönigsmann et al 1979). We used a new liquid preparation of 5-MOP (developed by Gerot Pharmazeutika) to extend earlier studies and to investigate the following issues: (a) the kinetics and range of the serum levels; (b) the phototoxic (erythemogenic) properties of the new preparation; (c) the therapeutic efficacy in psoriasis compared with that of 8-MOP; and (d) the spectrum and incidence of short-term side-effects. We included 169 patients in this study and assigned them randomly to treatment with either 0.6 or 1.2 mg/kg 5-MOP or 0.6 mg/kg 8-MOP. The distribution of skin types and kinds of psoriasis was almost identical in all three treatment groups.

Both psoralens were used as a liquid preparation in soft gelatine capsules and were given orally one (8-MOP) or two hours (5-MOP) before irradiation. Minimal phototoxicity dose (MPD) testing and treatment were performed according to the standard European protocol (Henseler et al 1981). The MPD served as the initial UVA treatment dose. Treatment consisted of four exposures per week until the complete clearing of psoriasis.

Pharmacokinetic studies revealed an absorption rate for 5-MOP of 25% that for 8-MOP. As expected from the serum level determinations, skin photosensitivity (as assessed by the MPD test) was dependent on the 5-MOP dosage. Thus the absence of phototoxicity reactions reported previously is not due to absent or reduced phototoxic properties but is presumably related to the lower serum levels of 5-MOP.

There were significant differences in the total UVA dose, number of exposures, and time needed for inducing complete remission of psoriasis between the low-dose 5-MOP group and the two other groups. Although the high-dose 5-MOP regimen also required significantly more exposures and a longer treatment period than the 8-MOP treatment, the UVA requirements were not substantially different. These data demonstrate that when 5-MOP is given at the same dose as 8-MOP, it is not as effective. However, if the dose of 5-MOP is doubled the therapeutic results resemble those obtained with 8-MOP. In particular, there is no significant difference in the most important parameter, namely the total UVA dose required for clearing of psoriasis.

Severe erythematous reactions (19%), nausea and vomiting (10%) and pruritus (17%) were frequently seen in the patients treated with 8-MOP. None of these side-effects was associated with the low-dose 5-MOP regimen. In the high-dose 5-MOP group, two patients (3%) had a severe phototoxic reaction, and another two experienced pruritus. Neither nausea nor vomiting was reported by any patient treated with 5-MOP.

5-MOP may be considered suitable for routine photochemotherapy for psoriasis, particularly for the treatment of fair-skinned, sun-sensitive patients and patients with 8-MOP intolerance (Tanew et al 1988a). However, because of their photochemical similarity, 5-MOP and 8-MOP may present similar cytogenetic hazards and risks of tumour formation (Zajdela & Bisagni 1981, Young et al 1983).

7-Methylpyridopsoralen

7-MPP is a recently synthesized monofunctional furocoumarin (Moron et al 1983). *In vitro* it has a high binding affinity for DNA and forms only monoadducts upon irradiation with UVA. 7-MPP is less mutagenic and three to four times less carcinogenic in the albino mouse than the bifunctional psoralens 8-MOP and 5-MOP (Dubertret et al 1985).

Because no data on systemic toxicity in man are available for 7-MPP, it was used topically. Treatments were given four times per week to six patients with chronic plaque-type psoriasis. 7-MPP and 8-MOP (10^{-2} M in a lipophilic excipient) were applied at different sites, and these were irradiated 20 minutes later. Both psoralens in this preparation induced complete remission in five patients and a marked improvement in the sixth patient after a mean of 15 exposures. To achieve this result, however, a much higher mean cumulative UVA dose was needed with 7-MPP than with 8-MOP. In contrast to 8-MOP, 7-MPP did not cause phototoxic reactions, even with high doses of UVA (over 25 J/cm^2) and exhibited only moderate melanogenic activity.

7-MPP seems to be highly effective in the topical photochemotherapy of psoriasis (Tanew et al 1988b). However, while not inducing acute side-effects, it requires considerably higher irradiation doses than 8-MOP. 7-MPP might represent a definite improvement in long-term safety because of its lower mutagenic and carcinogenic potential, but we do not yet know whether the higher UVA requirements will have some bearing in this respect.

4,6,4′-Trimethylangelicin

4,6,4′-TMA is another monofunctional furocoumarin that has been developed as a potential photochemotherapeutic agent for the treatment of psoriasis (Guiotto et al 1984). *In vitro* and preliminary clinical studies have shown that 4,6,4′-TMA has a strong antiproliferative activity and appears to lack phototoxicity (Cristofolini et al 1984).

To assess the therapeutic efficacy in psoriasis we followed a protocol similar to that used with 7-MPP. Again, no appropriate data exist on the systemic toxicity of 4,6,4′-TMA. Twelve patients with chronic plaque-type psoriasis were each treated with an 0.1% ethanolic solution of 4,6,4′-TMA and an 0.15% solution of 8-MOP in glycerol formal on separate areas of skin. Irradiation was performed 20 minutes after application of the furocoumarins four times weekly.

In six of the 12 patients both compounds induced complete remission or marked improvement of psoriasis. Five patients responded better to 8-MOP, whereas one patient showed complete clearance with 4,6,4′-TMA while exhibiting only moderate improvement in the area of skin treated with 8-MOP. As with 7-MPP, the UVA irradiation dose requirements for 4,6,4′-TMA were substantially higher than those for 8-MOP. Two patients showed a slight erythematous reaction to 4,6,4′-TMA treatment, whereas five patients showed a moderate or marked phototoxic response to the 8-MOP treatment. The melanogenic activities of both compounds were similar.

In summary, 4,6,4′-TMA has a definite antipsoriatic activity. However, the therapeutic efficacy is lower than that of 8-MOP, and the UVA dose requirements are higher (Tanew et al 1988b).

Use of other wavelengths (335 nm versus 365 nm)

Several photochemotherapeutic effects of 8-MOP (e.g. sunburn cell production) reach a maximum at irradiations with wavelengths between 320 and 335 nm but are not seen after 365 nm irradiations (Young & Magnus 1981). In the hope of improving oral PUVA therapy of psoriasis we compared the antipsoriatic efficacy of PUVA at these two wavelengths. The minimal phototoxicity dose (MPD) for both wavelengths was determined with a monochromator one hour after 8-MOP ingestion. Treatments were given according to the standard protocol four times weekly with monochromatic radiation of 335 and 365 nm in two template areas of psoriatic plaques. Three patients were treated in both areas with the same UVA dose, starting with the MPD at 335 nm (which was always lower than that at 365 nm). Three patients received doses adjusted to the relevant MPD for the wavelength used at each site. Healthy skin treated identically and diseased skin exposed to UVA without prior 8-MOP ingestion served as controls.

At equal UVA dosages based on the 335 nm MPD, psoriasis cleared completely with PUVA at 335 nm but not at 365 nm. Different UVA dosimetry based on the MPD of the wavelength used, 335 or 365 nm, resulted in remission with both wavelengths, but at 335 nm 50% less UVA was required for complete clearing than at 365 nm. Two patients showed cumulative phototoxic responses in both psoriatic and healthy areas treated with the MPD.

These results indicate that PUVA at 335 nm is more effective in clearing psoriasis with lower UVA doses (J. Brücke et al, unpublished observations 1989). This may contribute to greater long-term safety for photochemotherapy. On the other hand, Young et al (1988) reported a higher yield of tumours in hairless mice when 8-MOP was used with irradiation at wavelengths between 320 and 335 nm. Further studies are necessary to clarify this issue.

Conclusions

Photochemotherapy is a highly effective treatment for psoriasis and other skin diseases which has profoundly influenced dermatology, providing therapy for diverse disorders. In future new furocoumarin analogues with reduced cytogenetic hazards and less carcinogenic potential will be important. In addition, the development of photosensitizers unrelated to furocoumarins or the introduction of targeted drug delivery may open up new fields in the treatment of psoriasis.

Acknowledgements

We gratefully acknowledge the generous gift of 4,6,4′-methylangelicin by Dr F. Dall'Acqua, and of 7-methylpyridopsoralen by Drs L. Dubertret, D. Averbeck and E. Bisagni. 5-Methoxypsoralen was kindly provided by Gerot Pharmazeutika, Vienna.

References

Clinical cooperative study of PUVA-48 and PUVA-64 1979 Photochemotherapy of psoriasis. Arch Dermatol 115:576–579

Cristofolini M, Guiotto A, Rodighiero P et al 1984 Synthesis of new 6-methylangelicins as potential agents for the phototherapy of psoriasis. Acta Dermato-Venereol Suppl 113:170–172

Dubertret L, Averbeck D, Bisagni E et al 1985 Photochemotherapy using pyridopsoralens. Biochimie (Paris) 67:417–422

Epstein JK, Fukuyama K 1975 Effects of 8-methoxypsoralen (8-MOP) induced phototoxic effects on mammalian epidermal macromolecular synthesis in vivo. Photochem Photobiol 21:325–330

Fischer T, Alsins J 1976 Treatment of psoriasis with trioxsalen baths and dysprosium lamps. Acta Dermato-Venereol 56:383–390

Fritsch PO, Gschnait F, Kaaserer G et al 1979 PUVA suppresses the proliferative stimulus produced by stripping on hairless mice. J Invest Dermatol 73:188–190

Guiotto A, Rodighiero P, Manzini P et al 1984 6-Methylangelicins: a new series of potential photochemotherapeutic agents for the treatment of psoriasis. J Med Chem 27:959–967

Henseler T, Hönigsmann H, Wolff K, Christophers E 1981 Oral 8-methoxypsoralen photochemotherapy of psoriasis. The European PUVA study: a cooperative study among 18 European centres. Lancet 1:853–857

Henseler T, Christophers E, Hönigsmann H et al 1987 Skin tumors in the European PUVA Study. J Am Acad Dermatol 16:108–116

Hönigsmann H, Jaschke E, Gschnait F, Brenner W, Fritsch P, Wolff K 1979 5-Methoxypsoralen (Bergapten) in photochemotherapy of psoriasis. Br J Dermatol 101:369–378

Hönigsmann H, Jaschke E, Nitsche V, Brenner W, Rauschmeier W, Wolff K 1982 Serum levels of 8-methoxypsoralen in two different drug preparations: correlation with photosensitivity and UV-A dose requirements for photochemotherapy. J Invest Dermatol 79:233–236

Hönigsmann H, Wolff K, Fitzpatrick TB, Pathak MA, Parrish JA 1987 Oral photochemotherapy with psoralens and UVA (PUVA): principles and practice. In: Fitzpatrick TB et al (eds) Dermatology in general medicine. 3rd edn. McGraw-Hill, New York, p 1533–1558

Igali S, Bridges BM, Ashwood-Smith MJ, Scott BR 1970 Mutagenesis in *Escherichia coli*. Mutat Res 9:21–30

Melski JW, Tanenbaum L, Parrish JA, Fitzpatrick TB, Bleich HL 1977 Oral methoxsalen photochemotherapy for the treatment of psoriasis: a cooperative clinical trial. J Invest Dermatol 68:328–335

Moron J, Chi Hung N, Bisagni E 1983 Synthesis of 5H-furo(3′2′:6,7)benzopyrano(3,4-c) pyridin-5-ones and 8H-pyrano(3′,2′:5,6)benzofuro(3,2-c)pyridine-8-ones(Pyridopsoralens). J Chem Soc 225–229

Parrish JA, Fitzpatrick TB, Tanenbaum L, Pathak MA 1974 Photochemotherapy of psoriasis with oral methoxsalen and longwave ultraviolet light. N Engl J Med 291:1207–1211

Schenley RL, Hsie AW 1981 Interaction of 8-methoxypsoralen and near-UV light causes mutation and cytotoxicity in mammalian cells. Photochem Photobiol 33:179–185
Stern RS, Laird N, Melski J, Parrish JA, Fitzpatrick TB, Bleich HL 1984 Cutaneous squamous-cell carcinoma in patients treated with PUVA. N Engl J Med 310:1156–1161
Stern RS, Lange R 1988 Non-melanoma skin cancer occurring in patients treated with PUVA five to ten years after first treatment. J Invest Dermatol 91:120–124
Swanbeck G, Thyresson M 1974 Induction of respiration-deficient mutants in yeasts by psoralen and light. J Invest Dermatol 63:242–244
Tanew A, Ortel B, Rappersberger K, Hönigsmann H 1988a 5-Methoxypsoralen (Bergapten) for photochemotherapy. J Am Acad Dermatol 18:333–338
Tanew A, Ortel B, Hönigsmann H 1988b 5-methoxypsoralen and other new furocoumarins in the treatment of psoriasis. In: Douglas RH et al (eds) Light in biology and medicine. Plenum Publishing Corp, New York, vol 1:183–188
Turjanmaa K, Salo H, Reunala T 1985 Comparison of trioxsalen bath and oral methoxsalen PUVA in psoriasis. Acta Dermato-Venereol 65:86–88
Walter JF, Voorhees JJ, Kelsey WH, Duell EA 1973 Psoralen plus black light inhibits epidermal DNA synthesis. Arch Dermatol 107:861–865
Wolff K, Fitzpatrick TB, Parrish JA et al 1976 Photochemotherapy with orally administered methoxsalen. Arch Dermatol 112:943–950
Wolff K, Gschnait F, Hönigsmann H, Konrad K, Parrish JA, Fitzpatrick TB 1977 Phototesting and dosimetry for photochemotherapy. Br J Dermatol 96:1–10
Wolff K, Hönigsmann H 1981 Clinical aspects of photochemotherapy. Pharmacol Ther 12:381–418
Young AR, Magnus IA 1981 An action spectrum for 8-MOP induced sunburn cells in mammalian epidermis. Br J Dermatol 104:541–548
Young AR, Magnus IA, Davies AC, Smith NP 1983 Comparison of the phototumorigenic potential of 8-MOP and 5-MOP in hairless albino mice exposed to solar simulated radiation. Br J Dermatol 108:507–518
Young AR, Walker SL, Garmyn M 1988 A first approach to an action spectrum for 8-MOP phototumorigenesis in mice. J Invest Dermatol 90:175–178
Zajdela F, Bisagni E 1981 5-Methoxypsoralen, the melanogenic additive in suntan preparations, is tumorigenic in mice exposed to 365 nm UV radiation. Carcinogenesis 2:121–127

DISCUSSION

Truscott: Does the relative effectiveness of 335 and 365 nm light with 8-methoxypsoralen (8-MOP) simply reflect the extinction coefficients of the molecule at those two wavelengths?

Hönigsmann: Yes, 8-MOP does absorb better at the lower wavelength. It seems quite logical, but previously some people insisted that 365 nm was the more effective wavelength.

Carruth: How does a plaque change after treatment? What does the skin finally look like?

Hönigsmann: The goal is to abolish inflammation, scaling, and hyperproliferation and produce normal skin. Usually psoriatic plaques show some hyperpigmentation after treatment and the site of the plaque is visible for several months. Otherwise, the skin is normal.

Carruth: If the plaques recur, do they do so in the same areas?

Hönigsmann: Sometimes, but not necessarily.

Dougherty: Should we look for drugs that absorb further into the visible region of the spectrum? Would that affect the risk of carcinogenesis associated with this treatment? For example, photosensitizers are being screened for photodynamic therapy (PDT) that are very short acting, like 8-MOP, but that absorb much further into the visible wavelength regions.

Jori: The absorption wavelength isn't the critical factor. The mechanism by which the cell photodamage takes place is important. If we damage DNA, we still risk mutagenesis.

Hönigsmann: It depends whether we produce a lesion that can be repaired. It's ironic that the lesions that are repaired are more dangerous than those that are not.

Truscott: The use of 3-carbethoxy psoralen (3-CPs) was discontinued. What was the effect of treatment with this drug?

Hönigsmann: We stopped using it because it was practically ineffective against psoriasis. It did produce erythema in some patients.

Truscott: The singlet oxygen yield of 3-CPs is 600 times higher than that of 8-MOP.

Hönigsmann: That is a good indication that singlet oxygen production is not the decisive event in treatment.

Morgan: Was this only used in topical application?

Hönigsmann: Yes.

Dougherty: Is a DNA-active material necessary to treat psoriasis?

Hönigsmann: We don't know, but that's the common hypothesis. UVB treatment is quite effective against psoriasis and also makes DNA lesions. Thus DNA lesions appear to be necessary in treatment of psoriasis. Other types of therapy, for example with methotrexate, which is a cytotoxic drug and inhibits mitosis, are effective.

Henderson: Is the vascular supply to psoriatic lesions different from that to normal skin?

Hönigsmann: Yes; there are many more capillaries going into the upper layers of the dermis.

Henderson: Therefore it is similar to carcinoma-in-situ where the first changes are the vessels coming high up in the epithelial layer. It might be possible to exert a vascular effect rather than a cellular effect.

Morgan: How quickly is the vascular system built up?

Hönigsmann: The development of psoriatic lesions takes some weeks.

Dougherty: At what stage does carcinoma-in-situ become hypervascularized?

Kato: Bronchoscopy shows that carcinoma-in-situ has a high vascularity. The earlier stages of metaplasia are not so hypervascularized.

Dougherty: Is there a correlation between induction of vessels and conversion into carcinoma-in-situ?

Kato: It is difficult to say, but the proliferation of the cells needs vessels.

Jocham: We induced bladder tumours in rats by oral administration of butylbutanol nitrosamine (BBN). We can measure the density of the vascular system, even very small arterioles and venules. It takes about four months from the start of chemical induction of tumours before the appearance of something similar to carcinoma-in-situ, but the development of the vascular system starts much earlier. At the time when there is severe metaplasia one can already detect alterations in the vascular system. When carcinoma-in-situ first appears there is a much higher density of the small arterioles than in the normal bladder.

Lim: Is the induction of carcinoma-in-situ secondary to non-specific inflammation? Would administration of a compound that produces non-specific inflammation without necessarily being a carcinogen give the same effect?

Jocham: There are differences between the changes in the vascular system produced by general inflammation, such as that induced topically by *Escherichia coli*, and by carcinoma-in-situ. There is a higher density of vascularization in inflamed rather than normal tissue, and a much higher density in malignant tissue.

Jori: If we find a photosensitizer that acts on psoriatic plaques by vascular shutdown, the mechanism might be very different, perhaps leading to scar formation. Would this be cosmetically or ethically acceptable?

Dougherty: That is not inherent in the method. It should be possible to select the correct dose with any of these sensitizers.

Lim: One has also to find the appropriate light source. As yet there's no clinically available high intensity light source that can deliver adequate doses of light for PDT over a large area of skin surface.

Moan: Has the retention of cross-linked psoralen in the lesion been studied with radiolabelled drug?

Hönigsmann: Not in psoriatic skin, because we are no longer allowed to use radiolabelled compounds. Psoralens remain for a long time in cell culture. The cell survives for at least 14 days with a cross-link. In the skin it is shed; the cells enter the upper layers of the epidermis and are lost.

Moan: You suggested that new drugs cannot be given orally because of the possible toxic effects. Isn't there also a risk with new drugs that are applied topically? DMSO enters the circulation. That might also happen with some of these small molecules.

Hönigsmann: That is true for many drugs. If these are applied to only a few cm^2, it's unlikely that toxic levels are reached in the blood.

Greaves: Can those people who are treated with PUVA baths get systemic reactions?

Hönigsmann: Yes, they can.

Jori: It was mentioned that there is a general photosensitization. Did you see any side-effects?

Hönigsmann: This was only experimental work. After application of concentrated 8-MOP solution to about one third of the diseased body surface, patients had blood levels equal to those produced by oral administration.

Jori: When you give 8-MOP systemically and then irradiate the whole body, do you observe damage to normal skin?

Hönigsmann: We pre-determine the sensitivity of the individual and try to avoid the phototoxic skin reaction, but we don't always succeed.

Bonnett: Is PUVA treatment now generally available?

Greaves: It is in general use in Great Britain and in the United States.

Bonnett: If the treatment is done correctly is carcinogenesis no longer a worry?

Hönigsmann: The concerns are still there. That's why it is only used to treat severe cases.

Greaves: You can select out those patients who are particularly at risk, for example patients who are very sensitive to sunlight, those who have a history of skin cancer, and those who have had inorganic arsenical treatment. Our policy in Great Britain is that we don't treat people with psoriasis under the age of sixty, although we are becoming more confident about treating younger patients.

Carruth: What is the overall success rate of this treatment?

Hönigsmann: In about 80% of patients the skin returns to normal all over the body.

Jori: What is the interval between recurrences?

Hönigsmann: That varies for the individual patients.

Greaves: We did a trial for the Medical Research Council comparing topical treatment of psoriasis by dithranol with PUVA treatment (Rogers et al 1979). The relapse rate after we stopped the PUVA therapy was exactly the same—5.5 months before recurrence of psoriasis to half the level that it was at when the original treatment was started.

Dougherty: Is treatment then simply repeated?

Greaves: Most people continue with the maintenance treatment, but you could do it intermittently. The latter might be preferable because the total dosage may be slightly lower than if regular maintenance treatment is used.

Lim: The clearance with UVB therapy also lasts for 4–6 months.

Hönigsmann: If you compare UVB and PUVA treatment in the same patient, lesions always recur in the UVB-treated area two or three months before recurrence in the PUVA-treated area.

Reference

Rogers S, Marks J, Shuster S, Vella Briffa D, Warin A, Greaves M 1979 Comparison of photochemotherapy and dithranol in the treatment of chronic plaque psoriasis. Lancet i: 455–458

Extracorporeal photochemotherapy in the treatment of cutaneous T cell lymphoma and autoimmune disorders affecting the skin

Alain H. Rook, George T. Nahass, Barbara J. Macelis, William H. Macey and Stuart R. Lessin

Department of Dermatology, Hospital of the University of Pennsylvania, 3400 Spruce Street, Philadelphia, PA 19104, USA

Abstract. Extracorporeal photochemotherapy (ECPCT) is a new form of chemoimmunotherapy which involves the *ex vivo* exposure of malignant peripheral blood cells to 8-methoxypsoralen (8-MOP) and ultraviolet A (UVA) radiation followed by reinfusion of the treated cells. This treatment has resulted in an unprecedented number of prolonged remissions in patients with therapeutically resistant forms of cutaneous T cell lymphoma (CTCL) characterized by the systemic dissemination of a clonal population of malignant helper T lymphocytes. Although the mechanism of the beneficial effect is uncertain, an immune reaction to the reinfused modified T cells probably results in tumour regression. Because the T cell antigen receptor (TCR) is an immunogenic structure and because an identical TCR is present on the entire clonal population of malignant T cells in each individual with CTCL, modification of this structure represents the most likely target for the effects of 8-MOP and UVA. Understanding of the precise events leading to tumour regression in CTCL during treatment with ECPCT may lead to the expanded use of this therapy for other lethal haemopoietic malignancies.

1989 Photosensitizing Compounds: their Chemistry, Biology and Clinical Use. Wiley, Chichester (Ciba Foundation Symposium 146) p 171–182

Extracorporeal photochemotherapy has recently been approved in the USA as an effective therapy for certain advanced forms of cutaneous T cell lymphoma (CTCL). The treatment uses the natural compound 8-methoxypsoralen (8-MOP), which, when exposed to ultraviolet A (UVA), becomes a powerful chemotherapeutic agent. We shall review the results of this therapy and the theoretical mechanisms underying its efficacy. In addition, other potential uses of extracorporeal photochemotherapy will be discussed.

Extracorporeal photochemotherapy (also known as photopheresis) is a new form of chemotherapy which permits the extracorporeal exposure of pathogenic peripheral blood leucocytes to 8-methoxypsoralen (8-MOP) and UVA irradiation (Edelson et al 1987, Edelson 1988). The treatment uses a device which integrates an initial leukapheresis step with subsequent exposure of the separated leucocytes to UVA light within the same apparatus. The procedure is performed at a time following the oral ingestion of 8-MOP such that blood levels of 8-MOP are greater than 100 ng per ml at the initiation of the leukapheresis phase. The plasma and leucocyte collection, which contains 30–50% of the total peripheral blood leucocyte population, is then exposed to UVA (320–400 nm) for 270 minutes, which is sufficient to provide an average lymphocyte exposure of 2 J/cm^2. Following exposure of the cells to UVA, the treated cells and plasma are reinfused into the patient.

Clinical trials have demonstrated marked efficacy of extracorporeal photochemotherapy in the therapy of CTCL (Edelson et al 1987) which represents the most common type of adult T cell lymphoma (Winstock & Horm 1988). It consists of a proliferation of a clonal population of malignant $CD4^+$ helper T lymphocytes which develops initially within the skin (Lutzner et al 1975, Patterson & Edelson 1987). The Sézary syndrome is an advanced form of CTCL that is associated with the dissemination of the malignant T lymphocytes beyond the skin to the lymph nodes and to the peripheral blood, with a concomitant cutaneous erythroderma that is often disabling (Lutzner et al 1975). Profound defects in cell-mediated immunity are often associated with the Sézary syndrome (Edelson 1977). The clinical course of the Sézary form of CTCL is almost invariably one of relentless progression to death with a median survival of thirty months from the onset of disease, regardless of the type of treatment. The results of clinical trials which used extracorporeal photochemotherapy to treat the Sézary syndrome have been striking. At the University of Pennsylvania, eight of ten patients initially enrolled more than four years ago (in 1985) are alive and responding to therapy. It is particularly noteworthy that five of these patients initially developed complete clearing of malignant T cells from the skin, and four of these five are still alive and in complete remission. Overall, 24 of 29 patients in the national trial headed by Richard Edelson, currently at Yale University, exhibited marked clinical improvement.

Extracorporeal photochemotherapy represents a novel approach to chemotherapy because the treatments are performed *ex vivo* and thus eliminate many of the harmful and disabling effects of systemic chemotherapy. At our centre, the toxicity of this treatment has been minimal. The most serious side-effects have been: (a) rare episodes of hypovolemic hypotension occurring during the initial leukapheresis phase; and (b) the development of fever, which we presume to be immunologically mediated, four to twelve hours after therapy. Thus it should be emphasized that this therapy is well-tolerated and that patients undergoing extracorporeal photochemotherapy do not experience the usual toxic effects of chemotherapeutic drugs such as mucositis, alopecia, gastrointestinal upset and bone marrow suppression.

While the therapeutic efficacy of extracorporeal photochemotherapy for CTCL is convincing, the precise anti-neoplastic mechanisms of this treatment remain poorly defined. The *ex vivo* exposure of the neoplastic T lymphocytes to 8-MOP in the presence of UVA is thought to result in a critical modification of the malignant cell population such that, upon reinfusion, the host immune system is better able to recognize and eradicate the remaining tumour cells in the skin, lymph nodes, spleen, and other sites of T cell trafficking (Edelson 1988).

8-MOP is a member of the furocoumarin family of photosensitizing chemicals. Psoralens are known to intercalate within DNA in a dark reaction, but, when not exposed to light, psoralens are biologically inert. After exposure to UVA light psoralens form mono and bi-functional adducts with pyrimidine bases in the DNA. This results in covalent cross-linking of psoralen between base-paired strands (Scott et al 1976, Song & Tapley 1979). In addition to reactions at the DNA level, high affinity binding sites for psoralens have been identified on the surface of many mammalian cell types. Upon exposure to UVA, cell surface-bound psoralen becomes covalently linked to the membrane (Laskin et al 1985).

Either one or both of these biochemical reactions of psoralen are likely to be involved in the therapeutic procedure of extracorporeal photochemotherapy. It is known that the proliferative capacity of T lymphocytes is arrested following psoralen and UVA exposure (Edelson et al 1987). This is probably due to formation by psoralen of interstrand DNA covalent cross-links. Thus, one therapeutic mechanism in the treatment of CTCL involves the growth inhibition of the malignant T lymphocytes. However, results from another mode of treatment of the Sézary syndrome, namely leukapheresis (Pineda & Winkelmann 1981), suggest that mechanisms other than direct tumour cell killing by 8-MOP and UVA are responsible for the therapeutic effect of extracorporeal photochemotherapy. Leukapheresis alone, where peripheral blood leucocytes are removed and discarded, must be administered two or three times per week to produce disease stabilization or improvement. In contrast, extracorporeal photochemotherapy administered every four weeks is sufficient to achieve improvement in the majority of patients. Moreover, only a small fraction of the total *in vivo* pool of malignant lymphocytes is exposed to psoralen and UVA during each monthly treatment. Thus the majority of tumour cells in the skin, lymph nodes and spleen are not exposed to this treatment, indicating that another mechanism for tumour regression is being used.

Another possible mechanism for the beneficial therapeutic response to *ex vivo* psoralen and UVA treatments involves alterations of important cell surface determinants on the malignant T cells to render the cells more immunogenic. Khavari et al (1988), using the rat model of experimental allergic encephalitis (EAE), have established that the immunogenicity of T cell clones can be altered by exposure to psoralen and UVA. In this model, rats inoculated with myelin basic protein develop a paralytic illness associated with T cell destruction of the nervous system. The pathogenic T cells can be cloned *in vitro*. When these clones are administered to naive, syngeneic rats, EAE is reproduced. However, if the pathogenic clones are first inactivated by psoralen and UVA exposure, and then administered, the rats are protected from the development of disease upon subsequent inoculation with the pathogenic T cells. Control of the EAE-producing pathogenic T cells in rats that have first received the photoinactivated cells is probably mediated by psoralen- and UVA-induced suppressor T cells which have developed specifically in response to the modified cells. In related studies, Perez et al (1989) used a murine model of cutaneous allograft rejection and demonstrated that alloreactive T lymphocytes can also be rendered highly immunogenic following exposure to 8-MOP and UVA. Thus, in at least two different experimental models, the immunogenicity of T cells can be heightened by 8-MOP and UVA exposure.

The proposed cellular modification produced by psoralen and UVA could occur through psoralen-mediated changes at the DNA level resulting in altered gene expression for a cell surface molecule. Alternatively, UVA activation of psoralen at the cell membrane could result in changes in membrane proteins. Because only a fraction of the total *in vivo* malignant T cell pool is exposed to psoralen and UVA with each photochemotherapy treatment, and because complete tumour regression is obtained in some patients, one must implicate the alteration of a cell surface protein that represents a unique determinant for the entire *in vivo* clonal population of malignant T cells. Thus, when the *ex vivo*-treated peripheral lymphocytes are reinfused, an *in vivo* immune response is triggered against the entire pathogenic T cell clone. One candidate protein is the T cell antigen receptor (TCR) (Edelson et al 1987, Edelson 1988). The TCR is a well-characterized cell surface protein that is found on all peripheral T lymphocytes, and is responsible for the recognition of foreign antigens. Two types of TCRs have been identified on the surface of T cells. The $\alpha\beta$ TCR consists of two disulphide-linked polypeptides (designated α and β) and is expressed on the majority of T cells that recognize foreign antigens in the context of major histocompatibility complex molecules (Davis & Bjorkman 1988, Haskins et al 1983). A second type of TCR is composed of two disulphide-linked polypeptide chains designated γ and δ (Borst et al 1988). $\gamma\delta$ TCR is present on a small percentage (3%) of T cells. All four polypeptide chains consist of a variable (V), joining (J), and constant (C) region. (β and δ possess a diversity (D) region between the V and J regions.) The gene segments encoding the V,D,J, and C regions rearrange (by somatic recombination) during T cell development to generate diversity among receptor polypeptides. It is particularly noteworthy that a clonal population of T cells will process the identical T cell receptor on all its cells surfaces and will contain TCR genes that have all recombined in an identical manner. Thus, in any given individual with CTCL, all the malignant T cells will possess an identical TCR structure (Bertness et al 1985) making this the ‘fingerprint’ for all the tumour cells *in vivo*. It is also known that the variable region of the TCR is antigenic to the host immune system (Bertness et al 1985, Weiss et al 1985). Thus, 8-MOP plus UVA could mediate its effect through alteration of the TCR, which could lead to an improved ability of the host immune system to process and immunologically respond to the tumour cells.

A second hypothesis for the therapeutic efficacy of extracorporeal photochemotherapy is that patients with advanced forms of CTCL exhibit profound defects in multiple parameters of cell-mediated immunity, and a non-specific adjuvant effect on the immune system induced by the treatment of peripheral blood lymphocytes with 8-MOP and UVA could result in disease remission. Defects in cutaneous delayed type hypersensitivity (Nordquist & Kinney 1976), in *in vitro* T cell proliferative responses to microbial antigens (Edelson 1977), and importantly, marked defects in natural killer cell activity (Laroche & Kaiserlian 1983) are all associated with the Sézary syndrome. Because natural killer cell responses probably play a role in tumour surveillance, non-specific augmentation of this immune function, as well as other parameters of antitumour cell-mediated cytotoxicity, including macrophage killing and T cell cytotoxicity, could be significant. Indirect evidence that an immune adjuvant effect occurs is supported by clinical events. Approximately 50% of patients in the early stages of treatment exhibit high spiking fevers four to twelve hours after reinfusion of the irradiated cells. The fevers are not due to bacteremia and occur most frequently in patients who have had the most significant tumour regression (Heald & Edelson 1988). The temporal sequence of the fevers parallels the time course for release of the lymphokines tumour necrosis factor (TNF) and γ-interferon from macrophages and T cells, respectively. These lymphokines augment natural killer cell activity and also have inhibitory effects on tumour cell growth.

Patients also often show a transient but marked increase in cutaneous erythema and pruritus within twelve to twenty-four hours after reinfusion of the treated cells. Since the skin represents a substantial reservoir of lymphoma T cells, this clinical observation suggests that the treated cells may be trafficking to the skin and may be eliciting an antitumour response at that site.

Many of these hypotheses are now being tested in the laboratory. In addition numerous new clinical trials have been initiated using extracorporeal photochemotherapy to treat other neoplastic and autoimmune diseases. At our centre, early observations of a small number of patients with the autoimmune diseases progressive systemic sclerosis (PSS) and pemphigus vulgaris have demonstrated marked disease suppression during use of this therapy. Two patients with PSS have experienced skin softening, increased joint mobility, and reduced severity of Raynaud's phenomenon. Moreover, four individuals with pemphigus vulgaris have had a striking amelioration of their blistering disorder which could not be previously obtained with the use of high doses of glucocorticoids and immunosuppressive agents. Other disorders, including rheumatoid arthritis, multiple sclerosis, systemic lupus erythematosus and chronic lymphocytic leukaemia are also in the early phases of clinical trials.

The full spectrum of diseases that may respond to this exciting new form of therapy remains to be determined.

Much new information regarding the effects of extracorporeal photochemotherapy on the host immune response can be anticipated. The precise delineation of the mechanisms underlying the efficacy of this therapy will improve our therapeutic strategies for the appropriate treatment of other lethal haemopoietic disorders and T cell diseases.

References

Bertness V, Kirsch I, Hollis G, Johnson B, Bunn PA Jr. 1985 T cell gene rearrangements as clinical markers of human T cell lymphomas. N Engl J Med 313:534–538

Borst J, Van Dougen JJM, Bollhuis RLH et al 1988 Distinct molecular forms of human T cell receptor γ/δ detected on viable T cells by a monoclonal antibody. J Exp Med 167:1625–1644

Davis MM, Bjorkman PJ 1988 T cell antigen receptor genes and T cell recognition. Nature (Lond) 334:395–402

Edelson R 1977 Recent advances in the cutaneous T cell lymphomas. Bull Cancer (Paris) 64:209

Edelson R 1988 Light activated drugs. Sci Am 259:68–75

Edelson R, Berger C, Gasparro F et al 1987 Treatment of cutaneous T cell lymphoma by extracorporeal photochemotherapy: preliminary results. N Engl J Med 316:297–303

Haskins K, Kubo R, White J, Pigeon M, Kappler J, Marrack P 1983 The major histocompatibility complex-restricted antigen receptor on T cells. I. Isolation with a monoclonal antibody. J Exp Med 157:1149–1169

Heald P, Edelson RL 1988 Photopheresis for T cell mediated diseases. In: Callen J (ed) Advances in dermatology. Yearbook Medical Publishers p 25–40

Khavari PA, Edelson RL, Lider O, Gasparro FP, Weiner HL, Cohen IR 1988 Specific vaccination against photoinactivated cloned T cells. Clin Res 36:662

Laroche L, Kaiserlian D 1983 Decreased natural killer cell activity in cutaneous T cell lymphomas. N Engl J Med 308:101

Laskin JD, Lee E, Yurkow EJ, Laskin DL, Gallo MA 1985 A possible mechanism of psoralen phototoxicity not involving direct interaction with DNA. Proc Natl Acad Sci USA 82: 6158–6162

Lutzner M, Edelson R, Schein P, Green I, Kirkpatrick C, Ahmed A 1975 Cutaneous T cell lymphomas: the Sézary syndrome, mycosis fungoides and related disorders. Ann Intern Med 83:534–552

Nordquist BC, Kinney JP 1976 T and B cells and cell mediated immunity in mycosis fungoides. Cancer 37:714

Patterson JAK, Edelson RL 1987 Cutaneous T cell lymphoma and other leukemic and lymphomatous infiltrate of the skin. In: Fitzpatrick TB et al (eds) Dermatology in general medicine. McGraw Hill, New York p 1086–1118

Perez M, Edelson R, Laroche L, Berger C 1989 Specific suppression of antiallograft immunity by immunization with syngeneic photoinactivated effector lymphocytes. J Invest Dermatol 92:669–676

Pineda AA, Winkelmann RK 1981 Leukapheresis in the treatment of Sézary syndrome. J Am Acad Dermatol 5:544–549

Scott BR, Pathak MA, Mohn GR 1976 Molecular and genetic basis of furocoumarin reactions. Mutat Res 39:29–74

Song PS, Tapley KJ 1979 Photochemistry and photobiology of psoralens. Photochem Photobiol 29:1177–1197
Weiss LM, Hu E, Wood GS et al 1985 Clonal rearrangements of T cell receptor genes in mycosis fungoides and dermatopathic lymphadenopathy. N Engl J Med 313:539–544
Winstock MA, Horm JW 1988 Mycosis fungoides in the United States. JAMA (J Am Med Assoc) 260:306–308

DISCUSSION

Dougherty: Has extracorporeal photochemotherapy (ECPCT) been tried with AIDS patients?

Rook: Clinical trials will be done in San Francisco. The infectious particles of hepatitis B and non-A, non-B can be completely eliminated from transfused blood by exposure to psoralen and UVA. The infectivity of HIV and HTLV-1 within white cells can be abolished by pre-exposure to psoralen and ultraviolet light, without damaging the red blood cells.

Dougherty: Have you thought of combining this treatment with cutaneous exposure?

Rook: We have considered that. We are combining ECPCT with a variety of other therapies for patients who have incomplete responses to ECPCT alone. For example, we think an active immune response is necessary for the best results with this therapy. Therefore we have started additional treatment of some patients with leucocyte interferon, with some beneficial effect.

Dougherty: How about using interleukin 2?

Rook: One concern is that we might stimulate outgrowth of the malignant clone. This probably would not happen because it is not easy to grow the malignant clone in interleukin 2 alone. Moreover, patients who have advanced forms of cutaneous T cell lymphoma are not good LAK cell developers. We have also considered giving γ-interferon concomitantly. Other topical therapies are frequently used in these patients, including potent corticosteroids. We are also thinking about a variety of other treatments, such as oral retinoids.

Kessel: You treat the patient with oral psoralen and then irradiate the cells. Could you do the whole treatment *in vitro*?

Rook: The initial US Food and Drug Administration trial was designed to use the oral preparation. I understand that a soluble form of the drug which could be added directly to the extracorporeal cells will become available. However, additional clinical trials will be required for that new form of treatment.

Dougherty: Do all patients develop a febrile spike?

Rook: Many do. Those who ultimately have the best response to the therapy develop low grade fevers in the early phases of treatment. The patients with autoimmune disease subjectively feel warm, but they don't develop fever.

Wilson: Why does the irradiation take so long?

Rook: The 1 mm membrane cannot hold the large volume of cells that we need to treat. Therefore it takes a long time to recycle all the cells through it.

Wilson: What is the cell survival at the end of the procedure?

Rook: Immediately, by Trypan blue assay, the survival rate is high. However, when the treated cells are incubated *in vitro* with a lectin, the normal proliferative response is dramatically suppressed.

Wilson: You repeat the treatment many times, presumably on different fractions of the cells. Why do you think this effect isn't simply direct photobiological cell kill?

Rook: We only repeat the treatments once a month. Before this therapy was available, aggressive cases of Sézary syndrome were maintained in a stable state by leukapheresis and the malignant cells were discarded. That often had to be done once or twice a week.

Ullrich: Could you freeze your treated cells and use them as an immunogen to boost your patients between the extracorporeal photophoresis treatments? It appears that you are making these cells more immunogenic, and the patients need an immune response for the therapy to work. This might reduce the need for frequent photophoresis.

Rook: That's a very interesting idea. It has not been done. In *in vitro* mouse studies repeated immunization does appear to augment the immune response.

Hönigsmann: Why do we need the phototoxic reaction of psoralens and UVA? Could you leave out the psoralens? Might any other type of damage to the cells produce the same immune reaction?

Rook: The only way to determine that is by doing a sham study, which I think should have been done in the first place. We are going to do a trial for multiple sclerosis and a sham study will be done.

Dougherty: It would not be exactly a sham study. There is some evidence that UVB might produce similar effects.

Ullrich: Similar studies have been done with UVB in my department. Hostetter et al (1986) irradiated a non-antigenic spontaneous tumour cell line *in vitro* with UVB. The cells were rendered antigenic; they were rejected in normal mice where the non-antigenic parent would grow progressively. Cross-protection experiments showed that if mice are immunized with the antigenic variants they are protected against the growth of the spontaneous non-antigenic parental tumours (Hostetter et al 1989). That appears to be similar to the situation here.

Boon (1983) has obtained antigenic tumour variants with mutagens and Frost et al (1984) have used 5-azacytidine. It has been suggested that some kind of associative recognition system is involved. The antigens on the tumours are changed by the UVB. When you immunize with the highly immunogenic variant, the immune system recognizes the non-antigenic parent because there is some molecule on the cell surface of the parent that is not immunogenic but is antigenic. I don't know if the same thing would happen in extracorporeal photophoresis.

Rook: One criticism of the work on the cutaneous allograft model was that an important control was left out; the cells were not treated either with psoralen alone or with UVA alone. This has now been done and either one alone does not produce the same effect. The psoralen is needed to produce the suppressor cell response against the alloreactive T cells.

Dougherty: Has anyone tried the usual PUVA treatment for patients with T cell cutaneous lymphoma?

Rook: Yes; treatment failed for many of them. That's why their therapy was altered to extracorporeal photochemotherapy. In many of our patients the disease has already spread beyond the skin. PUVA therapy alone will be palliative, but the disease that's in the nodes, visceral organs and peripheral blood will continue to disseminate.

Dougherty: Why doesn't PUVA treatment induce the same immunological reaction as extracorporeal photochemotherapy?

Rook: From the animal studies we think this treatment requires an active immune response. By administering PUVA therapy directly and inducing destruction of the immune cells in the skin, one might abolish the ability to produce active immunization against a malignant T cell within the skin.

Moan: How prevalent is cutaneous T cell lymphoma?

Rook: There are estimated to be about a thousand new cases a year in the USA.

Carruth: What's the current thinking on the aetiology of multiple sclerosis that makes this treatment attractive?

Rook: One observes immune attack of the nervous system by autoreactive T cells. One sees similar phenomena in, for example, experimental allergic encephalitis. It has been reported that multiple sclerosis has a retroviral aetiology, but many people doubt the relationship of HTLV-1 infection to this disease. It's felt that by using similar treatment to that used for scleroderma and pemphigus vulgaris we might enhance suppressor cell responses against their cognate antigen.

Dougherty: Has the virus been isolated?

Rook: A group at the Wistar Institute in Philadelphia claims that using the polymerase chain reaction they have detected HTLV-1 in cells from the immune system of a small number of patients with multiple sclerosis.

Dougherty: Rheumatoid arthritis is presumably similar.

Kaye: I am fascinated that you intend to use extracorporeal photophoresis in a disease such as multiple sclerosis which has a chronic and unpredictable course. What are you going to measure? This is a crucial issue. Multiple sclerosis is a highly emotive condition and measurement of the results of any treatment is very difficult.

Rook: Besides intensive clinical examination by two independent, blinded observers, patients will undergo repeated magnetic resonance imaging (MRI) testing with gatilinium which apparently can accurately detect involved plaques within the central nervous system. This will be the first opportunity to determine what happens to involved plaques on a long-term basis with this treatment.

Kaye: Looking at clinical resolution is not useful because patients with multiple sclerosis go into remission naturally. Later they may get another plaque elsewhere.

Carruth: Do these plaques dissolve anywhere? If a neurological problem related to a plaque gets better, does the plaque disappear?

Kaye: Usually the opposite pertains; more plaques are shown by MRI than there are clinical lesions.

Carruth: But do the plaques disappear spontaneously, or do they remain when the neurological system compensates and recovers?

Kaye: I don't know.

Carruth: Have you got a baseline for the behaviour of plaques in untreated multiple sclerosis?

Rook: A control group will be given sham treatment and monitored in parallel with the treatment group.

Wilson: Do you use different drug and light doses for the various diseases?

Rook: We are using the same drug dose and light treatment for all the diseases that we are treating. The basis for embarking on some of the trials with autoimmune disease was the compassionate treatment of a limited number of patients who had explosive and potentially lethal scleroderma and explosive pemphigus. These patients had failed to respond to extremely high doses of standard therapy, but subsequently responded very quickly to extracorporeal photophoresis.

Wilson: Have Phase II dose escalation studies been done?

Rook: No.

Greaves: One must keep the problem of doing controlled clinical trials in perspective. For multiple sclerosis such trials are crucial because of the characteristic waxing and waning of the disease. But I was most impressed by the data on progressive systemic sclerosis (scleroderma), because that is an exorable progressive disease and patients do not normally undergo significant spontaneous remissions. In such a disease there is little advantage in doing controlled trials. In the other autoimmune connective diseases, such as rheumatoid arthritis and systemic lupus erythematosus, which tend to fluctuate more, you need to do placebo-controlled randomized trials.

Extracorporeal photochemotherapy is an extremely expensive treatment. Therefore instigation of a controlled clinical trial for, say, systemic lupus erythematosus would cost a great deal. It is also difficult to recruit enough patients for the treatment; one centre is unlikely to have enough patients who fulfil the protocol criteria.

Bown: How critical is the separation of the white and the red blood cells? Could we find a drug that doesn't influence the red cells? The most expensive element of the treatment is the cell separation.

Rook: It's very critical; as the haematocrit rises in the circulating compartments in the machine the efficiency of the treatment falls off rapidly. When the haematocrit is above eight percent the light does not penetrate to the white cells.

Bown: We could solve that problem by finding a photosensitizer that is activated by a wavelength which is not absorbed by haemoglobin.

Rook: You could just interchange the light cassette in a machine if you wanted to.

Bown: The light source is no problem.

Rook: What is happening to the cell membrane is possibly critical.

Dougherty: I suspect that there are many drugs of the type Dr Bown suggests. That would have a major impact on the treatment.

Bown: We can destroy leukaemic cells in bone marrow with phthalocyanines and red light, and normal stem cells survive (Singer 1988).

Ullrich: You don't want to destroy them; you want to alter them.

Bown: Yes, I know it's a long way from what we require for this therapy, but we can get an effect in that wavelength region.

Dougherty: What wavelength is used with merocyanine for bone marrow purging?

Bown: 540 nm. But we can use 675 nm with the phthalocyanines.

Kessel: There would be a lot of light scattering from the large ratio of red cells to white cells.

Bown: That doesn't matter; you would just use a thin film.

Kessel: You might break the red cells and release haemoglobin—that would lower the efficiency of photodamage.

Bown: It might be easier to separate that out than to separate red from white cells.

Rook: I am simply trying to find out why psoralen is working with this particular treatment!

Dougherty: How many research groups besides yours and R. L. Edelson's are addressing these problems?

Rook: Probably just a handful.

Ullrich: Do you know what kind of immune response is occurring against your treated cells? Is it a cytotoxic T cell (CTL) or a natural killer cell (NK) response?

Rook: We are studying that. Dr Brian Jegasothy has evidence that it is not a natural killer cell response. We are trying to clone the malignant T cell from the patients, but it is difficult and many groups have previously been unsuccessful. If we had the malignant T cell as a target we could examine what kind of immune responsiveness is generated. Is it a suppressor T cell that prevents its outgrowth? Is it a direct cytotoxic T cell? Is it release of tumour necrosis factor or interferon?

Dougherty: Is there no good animal model?

Rook: No. But one group has been able to establish the malignant T cells in mice that have severe combined immune deficiency. That may permit easier long-term growth of the malignant cell.

References

Boon T 1983 Antigenic tumor variants obtained with mutagens. Adv Cancer Res 39:121–151

Frost P, Liteplo R, Donaghue TP, Kerbel RS 1984 Selection of strongly immunogenic 'tum-' variants from tumors at high frequency using 5-azacytidine. J Exp Med 159:1491–1501

Hostetter LW, Ananthaswamy HN, Kripke ML 1986 Generation of tumor-specific transplantation antigens by UV radiation can occur independently of neoplastic transformation. J Immunol 137:2721–2725

Hostetter LW, Romerdahl CA, Kripke ML 1989 Specificity of antigens of UV radiation induced antigenic tumor cell variants measured *in vitro* and *in vivo*. Cancer Res 49:1207–1213

Singer CRJ, Linch DC, Bown SG, Huehns ER, Goldstone AH 1988 Differential phthalocyanine photosensitisation of acute myeblastic leukaemic progenitor cells: a potential purging technique for autologous bone marrow transplantation. Br J Haematol 68:417–422

Photodynamic therapy of early-stage lung cancer

Harubumi Kato, Norihiko Kawate, Komei Kinoshita, Hideki Yamamoto, Kinya Furukawa and Yoshihiro Hayata

Department of Surgery, Tokyo Medical College Hospital, 6-7-1 Nishishinjuku, Shinjuku-ku, Tokyo 160, Japan

Abstract. The clinical efficacy of photodynamic therapy (PDT) was evaluated in cases of early-stage lung cancer. Haematoporphyrin derivative was used as a photosensitizer and an argon dye laser and excimer dye laser were used as the light sources. A total of 199 lesions in 165 patients with lung cancer were treated by PDT. Forty-five lesions in 40 patients, which had been detected only by sputum cytology, were endoscopically suggestive of early-stage lung cancer. All patients showed normal chest X-ray findings. Thirty lesions in 26 patients were treated by PDT only, with or without other conservative treatment because of poor pulmonary function or for other reasons. All lesions initially showed complete remission. Recurrences were seen in three cases. Nine patients died of unrelated causes and one patient died after a recurrence. The other 16 patients are alive and apparently free of the disease. Fifteen lesions were treated surgically after PDT because the effectiveness of PDT was uncertain. Among these, five lesions showed complete remission histologically. Three of the PDT-only patients have now survived for more than five years. The longest surviving patient has gone eight years without apparent recurrence.

1989 Photosensitizing Compounds: their Chemistry, Biology and Clinical Use. Wiley, Chichester (Ciba Foundation Symposium 146) p 183–197

It has been known for many years that photosensitizing compounds show a photochemical reaction when exposed to light (Raab 1900). It has also been demonstrated that the photosensitizer haematoporphyrin derivative (HpD) has an affinity for malignant tissue (Lipson & Baldes 1960). The photochemical reactions in photodynamic therapy (PDT) include the production of singlet oxygen which has cytocidal properties and can therefore be used in the treatment of cancer. This therapy was pioneered by Dougherty in 1975 in the treatment of skin metastatic lesions of breast carcinoma (Dougherty et al 1978). Kato et al (1981) and Hayata et al (1983) began to investigate this technique in 1978 and demonstrated the therapeutic effectiveness in canine tumour models, including, for the first time, early-stage lung cancer. We began to apply this method to clinical cases in 1980 (Hayata et al 1982), being the first group to treat early-stage cancer of the lung (or of any organ) by PDT. The patient died of another disease four years later, but was shown at autopsy to be free of malignant disease. We now have three patients who have survived for five years, including one who is the first person to have survived so long after treatment by PDT alone (Kato et al 1986a). In this paper we describe the application of photodynamic therapy to lung cancer, especially early-stage cases.

Materials and methods

Photosensitizers. The photosensitizer used was haematoporphyrin derivative (HpD), supplied as: Photofrin I (Photofrin Medical Inc., Cheektowaga, NJ, USA); Photofrin II, (Photomedica Inc., Raritan, NJ, USA); and HpD (Seikagaku Kogyo Co. Ltd, Chiba, Japan).

Therapeutic equipment. An argon dye laser [Spectra-Physics (model 171-08, 375-03, USA), Cooper LaserSonics (770D) and Aurora (USA) and Fujinon PDT (Fujishashinkoki Co. Ltd, Omiya, Japan)] and an excimer dye laser [Hamamatsu Photonics Co. Ltd (Hamamatsu, Japan)] were used for this study. Quartz fibres (400 μm) were used to introduce the laser beam to the lesions through the instrumentation channel of the fibre-optic bronchoscope (Olympus Co. Ltd, Tokyo, Fujinon, Fujishashinkoki Co. Ltd, Omiya, Japan).

Therapeutic methodology

PDT was performed 72 hours or more after intravenous injection of between 2.5 and 5.0 mg/kg body weight Photofrin I, or between 1.5 and 2.5 mg/kg body weight Photofrin II. Superficial and/or interstitial photoradiation was performed, depending on the nature of the lesion. For superficial PDT, a beam with a power of about 200 mW/cm^2 was administered for approximately 20 minutes. The estimated energy dose (e.e.d.) which the lesion received in such cases was therefore about 240 J/cm^2. For interstitial irradiation, a light beam with a power of around 100 mW at the fibre tip was used.

Evaluation of efficacy

We have been evaluating tumour response to PDT endoscopically, histologically and cytologically at one month after treatment for inoperable cases. In surgically resected or autopsied cases, the treated areas were examined macroscopically and histologically. Tumour response was divided into four categories: complete remission, significant remission, partial remission and no remission. Complete remission means no endoscopic, macroscopic, cytological or histological evidence of a tumour. Significant remission means that 60% or more of the original tumour volume has disappeared, as measured by endoscopy or macroscopically, and partial remission means that less than 60% but more than 20% of the tumour has disappeared, as observed by endoscopy or macroscopically. No remission means that less than 20% has disappeared.

Lung cancer cases

A total of 199 lesions in 165 patients with lung cancer were treated by PDT. Of these, 45 lesions in 40 patients examined by endoscopy were suggestive of early-stage lung cancer. The age distribution among these early-stage patients ranged from 36 to 82 years (average 65.6 years), and there were 38 males and one female. Histologically, 44 patients had squamous cell carcinoma and one patient had adenocarcinoma. Endoscopically, 35 lesions were finely granular and/or of the superficial type, and ten lesions were of the protrusion type. The areas of the lesions ranged from 1×0.5 to 5.3×2 cm, and the volumes from $0.2 \times 0.2 \times 0.2$ to $1 \times 0.5 \times 0.5$ cm. For 26 of these patients with early-stage lung cancer, surgery was not possible because of poor pulmonary function, possible cardiac failure or refusal to consent to the operation.

Results

Complete remission was obtained in 30.15% of the 199 lesions, significant remission in 53.26%, partial remission in 15.08% and no remission in 1.51%. In advanced lesions, opening of bronchi previously obstructed by tumours was obtained for 53 out of 72 lesions (73.61%).

Among 45 early-stage lesions (40 patients) complete remission was obtained in 35 lesions (77.8%). Table 1 shows the details of 26 patients (30 lesions) for whom surgery was not possible. For 12 of these lesions PDT could not be expected to succeed because of difficulties at the anatomical site or problems with endoscopic observation of the extent of the lesion. Therefore these lesions were treated by additional radiotherapy and by oral chemotherapy to prevent recurrence. Apparent complete remission was observed in all of these, but recurrence was later recognized endoscopically in two lesions (16.7%). Detailed examination of autopsied specimens revealed recurrence in one other case. The recurrent cases were treated by surgery and/or radiotherapy. The remaining 18 lesions, for which the post-treatment evaluation was 'probably curative', showed complete remission, but recurrence was observed in one lesion (5.5%). Ten patients died at 46, 34, 33, 32, 31, 30, 23, 23(10), 16 and 16 months after PDT as a result of other diseases, except for one case of recurrence of lung cancer at 14 months.

Figure 1 shows fibre-optic bronchoscope findings for the longest reported surviving patient with early-stage lung cancer treated by PDT. This patient was a 59-year-old female who suffered a productive cough for six months. Her chest X-ray film was negative but sputum cytology showed squamous cell carcinoma.

TABLE 1 Results of PDT in inoperable early-stage lung cancer patients[a]

Case no.[b]	*Age*	*Location of the tumour*	*Reasons for PDT only*
Group I			
1. J. M.	74	right B^{2}_{b} bronchus	Refused surgery
2. K. O.	76	left B^{10}_{a} bronchus	Poor pulmonary function
3. T. O.	59	right upper lobe bronchus	Poor pulmonary function
4. S. K.	70	right upper lobe bronchus	Poor pulmonary function
5. M. C.	62	left B^{1+2}_{ab} bronchus	Poor pulmonary function
6. T. K.	70	right upper lobe bronchus	Poor pulmonary function
7. Y. K.	62	left B^{1+2} bronchus	Refused surgery
	67	left B^{6} bronchus	Refused surgery
	67	right B^{2} bronchus	Refused surgery
8. E. A.	71	left B^{3} bronchus	Refused surgery
9. S. H.	70	right upper lobe bronchus	Refused surgery
10. T. O.	75	left B^{1+2} bronchus	Poor pulmonary function
11. M. S.	77	left lower lobe bronchus	Poor pulmonary function
12. K. T.	44	left upper lobe bronchus	Refused surgery
13. K. T.	79	left B^{6} bronchus	Poor pulmonary function
14. C. S.	62	right B^{1} bronchus	Poor pulmonary function
15. K. Y.	59	right B^{3}_{ab} bronchus	Poor pulmonary function
16. K. C.	65	right upper lobe bronchus	Refused surgery
17. M. F.	60	left B^{3} bronchus	Poor pulmonary function
18. B. T.	64	right $B^{1,2}$ bronchus	Poor pulmonary function
19. I. K.	75	right upper lobe bronchus	Poor pulmonary function
	75	trachea	Poor pulmonary function
20. K. A.	63	left upper lobe bronchus	Refused surgery
21. I. Y.	65	right B^{6} bronchus	Refused surgery
Group II			
22. T. U.	58	right B^{7} bronchus	Refused surgery
23. Y. S.	81	left lower lobe bronchus	Poor pulmonary function
	82	right B^{7} bronchus	Poor pulmonary function
24. T. K.	58	left B^{9} bronchus	Refused surgery
25. T. K.	41	left upper division bronchus	Refused surgery
26. K. M.	79	left B^{10}_{a} bronchus	Poor pulmonary function

[a]All patients, except no. 3, were male, and all lesions had squamous cell carcinoma histology, except that of patient no. 11 which was adenocarcinoma.
[b]Group I, biopsy(+), cytology(+); Group II, cytology(+).
[c]C, probably curative; N, probably non-curative.
[d]Pneumonia or chronic obstructive pulmonary diseases.
[e]Cerebral infarction.
[f]Prostatic carcinoma.
[g]Liver carcinoma.
Rec, recurrence.

Post-PDT evaluation[c]	*Combined therapy*	*Survival (months)*	*Prognosis*
C		46	Dead[d]
N	Radiation and Immunotherapy	31	Dead[e]
C		95	Alive
C		33	Dead[f]
C		16	Dead[d] Rec.
C		81	Alive
N	Chemotherapy and	80	Alive
N	Radiation	21	
C		16	
N	Chemotherapy	23	Dead[d] Rec.
N	Chemotherapy	46	Alive
N	Radiation	32	Dead Rec.
C		30	Dead
C		38	Alive
N	Radiation	34	Dead
C		26	Alive
C		26	Alive
C		23	Alive
C		20	Alive
N	Radiation	12	Alive
N	Radiation	13	Alive
N	YAG laser	13	
C		10	Alive
C		5	Alive
C		57	Alive
N	Chemotherapy	23	Dead[d]
N	Chemotherapy	10	
C		52	Alive
C		48	Alive
C		16	Dead[g]

TABLE 2 Results of PDT in early-stage lung cancer cases resected after PDT[a]

Case no.[b]	*Age*	*Location of the tumour*	*Result*[c]	*Post-PDT evaluation*[d]	*Combined therapy*	*Survival (months)*	*Prognosis*
Group I							
1. Y. A.	50	left upper lobe bronchus	SR	N	Chemotherapy	95	Alive
2. T. F.	54	right B^8 bronchus	SR	N	Chemotherapy	84	Alive
3. S. N.	74	left B^{1+2} bronchus	CR	N	Immunotherapy	82	Alive
4. B. N.	70	left upper lobe bronchus	SR	N	Radiation	41	Dead
5. K. S.	62	left B^{1+2}_{ab} bronchus	CR	N		54	Alive
6. T. M.	81	right middle lobe bronchus	SR	N		52	Alive
7. K. N.	65	right B^3bi	SR	N		57	Alive
8. Y. O.	50	right B^{10} bronchus	SR	N		45	Alive
9. K. O.	75	left upper lobe bronchus	SR	N	Immunotherapy	23	Dead Rec.[e]
10. S. K.	64	left B^3 bronchus	CR	C		34	Alive
Group II							
11. T. I.	70	right B^2_a bronchus	SR	N		58	Alive
12. S. M.	36	right upper lobe bronchus	CR	C		55	Alive
13. N. Y.	54	right B^6 bronchus	SR	N		11	Dead
14. B. T.	62	left main bronchus	CR	C		37	Alive
15. T. S.	67	left upper lobe bronchus	SR	N		8	Alive

[a]All patients were male and all lesions were of squamous cell carcinoma histology.
[b]Group I, biopsy(+), cytology(+); Group II, cytology(+).
[c]SR, significant remission; CR, complete remission.
[d]C, probably curative; N, probably non-curative.
[e]Recurrence at left main bronchus (invasion from lymph node).

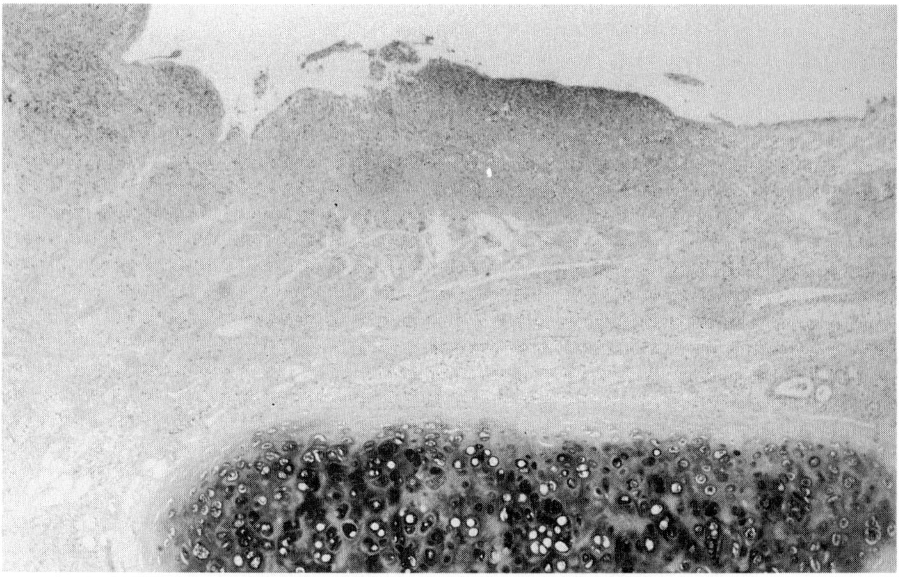

FIG. 4. Histological findings of the resected specimen after PDT for a 74-year-old male. Complete necrosis of tumour tissue is observed.

who had a superficial early squamous cell carcinoma in the left B^{1+2} bronchus. The tumour was found to have completely disappeared. His chest X-ray film showed normal findings.

Table 3 shows the results of PDT in early-stage lung cancer cases according to the location of the tumour. Complete remission was obtained for 35 lesions which were located in only two or fewer bronchial branches. Tumours recurred for four of these lesions. Those lesions which showed incomplete remission were extensive, involving more than two branches and/or sub-subsegmental or more peripheral bronchi.

Complications

Several complications were experienced after PDT for lung cancer, such as sunburn due to sensitivity to light (90% of patients), obstructive pneumonia (5%), perforation of the bronchial wall (2%) and massive bleeding (1%). Obstructive pneumonia occurred because of an increase in secretions. Perforation occurred when advanced tumours invaded the entire bronchial wall from beyond the bronchus. One patient with an advanced tumour died from massive bleeding four months after ionizing radiotherapy following PDT. The actual cause of death was unknown.

TABLE 3 Results of PDT in early-stage lung cancer patients according to the location of the tumours (45 lesions in 40 patients)

<table>
<tr><th></th><th>Main bronchus</th><th>Lobar bronchus</th><th>Segmental bronchus</th><th>Subsegmental bronchus</th><th>Sub-subsegmental bronchus</th><th>More peripheral bronchi</th><th>No. of lesions</th></tr>
<tr><td rowspan="3">Complete remission</td><td>2</td><td>12</td><td>11</td><td>6</td><td></td><td></td><td rowspan="3">35</td></tr>
<tr><td></td><td></td><td colspan="2">3</td><td></td><td></td></tr>
<tr><td></td><td colspan="2">1</td><td></td><td></td><td></td></tr>
<tr><td rowspan="5">Tumour nests remained</td><td></td><td>1</td><td>1</td><td></td><td></td><td></td><td rowspan="5">10</td></tr>
<tr><td></td><td colspan="3">1</td><td></td><td></td></tr>
<tr><td colspan="3">2</td><td colspan="3">1</td></tr>
<tr><td></td><td colspan="2">2</td><td></td><td></td><td></td></tr>
<tr><td></td><td></td><td colspan="2">1</td><td colspan="2">1</td></tr>
<tr><td rowspan="2">Recurrence</td><td></td><td></td><td colspan="3">2</td><td></td><td rowspan="2">4</td></tr>
<tr><td colspan="3">1</td><td colspan="3">1</td></tr>
</table>

Discussion

Haematoporphyrin derivative has a greater affinity for and is retained longer in malignant tissue than normal tissue. HpD is excited by light with wavelengths from the ultraviolet to red. The most effective excitation wavelength is at the Soret band, 405 nm. There are some Q bands at 488, 514 and 630 nm. Therefore a wavelength of 405 nm is usually used to elicit HpD fluorescence for diagnostic applications (Kato & Cortese 1985), whereas 630 nm wavelength light is used for cancer treatment because it penetrates tissue most deeply (Otawa et al 1982). In this study argon dye lasers and an excimer dye laser were used. The excimer dye laser system was newly developed for photodynamic diagnosis and therapy by the authors. The laser beam is obtained by excitation of XeCl gas. Because the wavelength of the light produced is 308 nm, rhodamine solution is used to tune to it 630 nm. The mechanisms of cancer treatment have been previously postulated (Weishaupt et al 1976, Sakai et al 1985). Singlet oxygen is thought to play an important role in the cytocidal effects. We have shown earlier that HpD is absorbed by the cytoplasm, especially in mitochondria (Saito et al 1985), and that the first cellular changes begin in mitochondria (Kato et al 1986a). Berns et al (1982) described similar results. The cytocidal effects are thought to reflect damage to the tricarboxylic acid cycle.

This paper evaluates the effectiveness of photodynamic therapy in early-stage lung cancer. All the patients showed normal chest X-ray findings but sputum cytology examination played an important role in the detection of malignant diseases. We treated 40 patients with early-stage lung cancer (45 lesions) with PDT. Complete remission was obtained for 35 of the lesions (77.8%). For all 18 'inoperable' lesions for which PDT was evaluated as 'probably curative' apparent complete remission was obtained (100%), and recurrence was later recognized in only one lesion (5.6%). This emphasizes that this new therapy offers the possibility of curing certain early-stage lung cancer lesions with a single treatment.

For those patients for whom complete remission was not obtained, the reasons were investigated by looking at the histology of resected specimens. Complete remission was not obtained for lesions which were anatomically difficult to irradiate or located submucosally, if irradiation from an angle of 90° was impossible, for lesions located beyond the cartilage or for extensive lesions. Improvements in the laser equipment, quartz fibres, and endoscopes are necessary for such cases. For successful PDT in early-stage lung cancer with the available technology: the lesion should be visible endoscopically; it should be possible to recognize the peripheral margin of the tumour; and submucosal tumour invasion should be limited to within the bronchial cartilage.

Patients in this study experienced several complications, such as sunburn obstructive pneumonia, perforation of the bronchial wall and massive bleeding. Therefore, they were told to avoid sunlight for at least three weeks after PDT, bronchial toilet was performed daily for three days after PDT, and light dosage should be limited in large tumours extending beyond the walls of the organs. PDT holds great potential for curing early-stage cancer, for palliative treatment of advanced-stage cancer and local improvement of lesions, for combination therapy with surgery (Kato et al 1985) and with ionizing radiation and chemotherapy. More stable, definitive and successful results should be obtained when drugs with more equal distribution in the tumour tissue are found and when deeper tissue penetration by longer wavelength light is achieved.

Acknowledgements

This study was supported in part by a Grant-in-Aid for scientific research from the Ministry of Education and a cancer research grant from the Ministry of Health and Welfare of Japan. We should like to thank Associate Professor J. Patrick Barron of St. Marianna University School of Medicine for his review of the manuscript.

References

Berns MW, Dahlman A, Johnson FM et al 1982 In vitro cellular effects of hematoporphyrin derivative. Cancer Res 42:2325–2329

Dougherty TJ, Lawrence G, Kaufman JE et al 1978 Photoradiation in the treatment of recurrent breast carcinoma. J Natl Cancer Inst 62:231–237

Hayata Y, Kato H, Konaka C et al 1982 Hematoporphyrin derivative and laser photoradiation in the treatment of lung cancer. Chest 81:269–277

Hayata Y, Kato H, Konaka C et al 1983 Fiberoptic bronchoscopic photoradiation in experimentally induced canine lung cancer. Cancer (Phila) 51:50–56

Kato H, Cortese DA 1985 Early detection of lung cancer by means of hematoporphyrin derivative fluorescence and laser photoradiation. Clin Chest Med 6:237–253

Kato H, Konaka C, Ono J et al 1981 Cancer localization by detection of fluorescence by means of HpD administration and krypton ion laser photoradiation in canine lung cancer. Lung Cancer 21:439–445

Kato H, Konaka C, Ono J et al 1985 Preoperative laser photodynamic therapy in combination with operation in lung cancer. J Thorac Cardiovasc Surg 90:420–429

Kato H, Konaka C, Kawate N et al 1986a Five-year disease-free survival of a lung cancer patient treated only by photodynamic therapy. Chest 90:768–770

Kato H, Aizawa K, Shinohara H et al 1986b Cytomorphological changes caused by hematoporphyrin derivative and photodynamic therapy. Lasers in the Life Sciences 1:13–27

Lipson RL, Baldes EJ 1960 The photodynamic properties of a particular hematoporphyrin derivative. Arch Dermatol 82:508–516

Otawa M, Aizawa K, Kato H et al 1982 Evaluation of tissue penetration of laser irradiation. Gendai-No-Shinryo 24:25–29

Raab O 1900 Uber die Wirkung fluorescierender Stoffe auf Infusorien. Z Biol 39:524–546

Saito M, Kato H, Konaka C et al 1985 Affinity of HpD for mitochondria. Igaku-No-Ayumi 132:600–602

Sakai H, Aizawa K, Kato H et al 1985 Detection of singlet oxygen from HpD using luminol. J Tokyo Med Coll 43:940–949

Weishaupt KR, Gomer CJ, Dougherty TJ 1976 Identification of singlet oxygen as the cytotoxic agent in photoinactivation of a murine tumor. Cancer Res 36:2326–2329

Kato: From the results described, about 90% of early cases
stage lung cancer is not the same as asymptomatic lung
peripheral-type, early-stage lung cancers are asymptomatic. E
type lung cancer, especially squamous cell carcinoma, is symp
curable. Of peripheral tumours which are less than 2 cm in
metastasized, 84% are curable.

Jori: You mentioned that one reason for failure or limit
might be the non-homogeneous distribution of HpD in tum
have any evidence for this?

Kato: We injected HpD into a mouse tumour and found th
was not equally distributed.

Jori: Which compartments of the tumour accumulate the
fluorescent material?

Kato: The interstitial connective tissue and the tumour cells
fluorescence. The necrotic part does not fluoresce as stron

Henderson: Dr Kato, could you have treated these lesions

Kato: YAG lasers are very powerful and there is a danger
bronchial walls. Perhaps if the power were decreased it m
treat superficial tumours.

Bown: We've treated small tumours with a YAG laser, but
alongside much larger tumours which were the main clinical pr
to see if we could treat the small areas as well.

Carruth: Can you treat the tumour with benefit to the

Bown: Not so far, because the treatment is not well en
match tissue necrosis with the extent of the tumour being tr
be possible in the future.

Dougherty: For small tumours it is desirable to aim treatm
wall, but use of a YAG laser carries the danger of light penetr
wall.

Kato: Also, in the very early stages of lung cancer the b
the tumour and normal tissue is unclear. With the YAG las
irradiate the whole area, including the normal tissue. Argon-
do not have this problem.

Jocham: In our preliminary experience of treating six
cancer, we found clear indications where one couldn't use th
is certainly indicated in some types of lung cancer.

DISCUSSION

Dougherty: Dr Kato, you have used a wide range of light doses for treatment. Has that affected your results?

Kato: Initially, more than five years ago, we tried to treat lung cancer at every stage of the disease. To ensure effective therapy we used high doses. Now we know more about dosages, and we use about 200 J/cm^2.

Dougherty: You are still using an uncut fibre. Would you get a better light distribution using a lens-type fibre?

Kato: Yes, I hope to use a cylindrical fibre or lens-type fibre for extensive lesions in the future. But if we use these fibres we shall need a more powerful laser because the light density will be less.

Dougherty: The new fibres transmit at least 80% of the power from the laser, so that's not a problem. Also, only a small amount of power is required to treat these tumours.

Kato: The tumours that we usually treat are very small. With a cylindrical fibre most of the light goes in the wrong direction.

Dougherty: I am not suggesting that you use a cylindrical fibre to treat small lesions. The lens-type fibre is best for those.

Berenbaum: Did the high light doses, up to 800 mW/cm^2, produce heating effects?

Kato: We now use powers of less than 200 mW/cm^2, and there are no heating effects. But in the early treatments at 800 mW/cm^2 there was probably some heating.

Carruth: In which patients do you give chemotherapy or radiotherapy in addition to PDT?

Kato: Nine patients (12 lesions) were additionally treated by radiotherapy and chemotherapy. We did this when we could not be sure that the whole tumour could be treated by PDT, either because the lesion was not visible or because the tumour was difficult to irradiate. The latter cases include widespread tumours and those at certain anatomical sites, such as the medial wall of the left upper division bronchus and the right upper lobe bronchus in the direction of B_1 and B_2, where we couldn't evaluate the results endoscopically.

Dougherty: A cylindrical fibre might be useful for these cases.

Bown: Surely the clinical problem now is one of diagnosis not therapy? We know this treatment works for small tumours; the problem is knowing exactly how far the tumours extend.

Dougherty: For example, one tumour that you described was diagnosed as early-stage lung cancer but was almost totally obstructing the bronchus. How did you judge that to be early stage?

Kato: Our criteria for diagnosing early-stage lung cancer are that the tumour should be localized within the bronchial wall and there should be no metastasis. It is always difficult to be precise about metastasis.

Dougherty: Most of your patients had not previous
Wouldn't it be a good idea to give them all radiotl
the treatment is not complete? Is there a reason fo
Kato: We used additional therapy for nine patients
above. For the other patients with small visible lesions
to be sufficient.
Hönigsmann: Does the histological type of the tun
of PDT? You have shown us many cases of squam
about other types?
Kato: We treated adenocarcinoma, small cell c
carcinoma. But those were advanced cases. Squamou
in the epithelium of the bronchial mucosa, wherea
submucosally and is not suitable for treatment by P
small cell and large cell carcinomas.
Carruth: Diagnosis of asymptomatic bronchial car
you tell us about your sputum cytology screening p
Kato: In Japan we have several kinds of lung cancer
for the detection of lung cancer was started two year
are males and females over the age of 40 and
examinations every year. For financial reasons, sput
for males over 50 years old. Screening for lung can
private companies. Lung cancer studies are publicized
journals. Sputum cytology can be very effective in det
type lung cancers that are not visible by X-ray e
successful for these tumours; the five-year survival
Dougherty: Are these studies done for smokers c
Kato: We look at smokers, night workers and an
detection rate for lung cancer is generally about 30 pe
cytology gives a detection rate of 60 to 70 per 100 000 p
this depends upon the groups studied. The rate is hi
the studies by private companies the subjects are you
is low.
Lim: What are the false positive and false negative
Kato: There is a large problem here. When sputum
cell carcinoma it is often impossible to localize
bronchoscopy because the disease is at an early st
particularly the peripheral ones, are visible. In thos
repeat sputum cytology and chest X-ray every thre
people show obvious cancer by X-ray or fibre-optic
years later.
Berenbaum: How much does it cost to detect lu
Kato: It's very expensive: £20 000–24 000 to dete
Berenbaum: What percentage of those detected

Long-term experience with integral photodynamic therapy of TIS bladder carcinoma

D. Jocham, M. Beer, *R. Baumgartner, G. Staehler, *E. Unsöld

*Department of Urology of Ludwig-Maximilians-Universität, Klinikum Grosshadern, Munich and *Central Laser Laboratory, GSF Neuherberg, West Germany*

Abstract. Fifteen patients in whom superficial bladder tumours (TIS GII-III) recurred after the unsuccessful application of other treatments were treated by photodynamic therapy (PDT) and observed for 24–54 months. The results of therapy were evaluated by regular three-monthly check-ups (endoscopy, cytology, bladder mapping, renal ultrasonography) as well as by computed tomography (CT) examination at 8- to 13-month intervals. In nine patients treated by PDT no tumour recurrence has been found over the whole observation period. Two patients have remained free of tumour (12 and 14 months) after repeated transurethral resection (TUR) and Nd-YAG laser therapy after PDT. Two patients required a second application of PDT before the treatment goal was achieved. In one of these cases a circumscribed dysplasia which appeared at the left ostium 26 months after PDT was treated by TUR and Nd-YAG laser. Further observations showed it had been successfully removed. In six patients slight mucosal atypia persisted for at least 2.5 years. In the short-term course of further cases ($n = 12$) one cystectomy has to be performed because of bladder shrinkage. According to these preliminary results, PDT (with strict patient selection—worst-case situation with recommended cystectomy) is justified for recurrent superficial TIS bladder carcinoma.

1989 Photosensitizing Compounds: their Chemistry, Biology and Clinical Use. Wiley, Chichester (Ciba Foundation Symposium 146) p 198–208

One reason for tumour development is thought to be diffuse sprinkling of the bladder wall with carcinogenic components of urine. Almost 80% of recurrent bladder carcinomas show multifocal growth within the bladder which cannot be detected by routine endoscopy. These foci of dysplasia and carcinoma-in-situ can represent the reservoir for rapidly recurrent tumours (Althausen et al 1976). Therefore homogeneous irradiation of the bladder using photodynamically efficient light in the correct wavelength range for photodynamic therapy (PDT) is necessary for detection of all tumour areas present in the bladder (Jocham et al 1981). The aim of treatment is complete tumour detection (Jocham et al 1987) and destruction, at least at the time of irradiation (Baghdassarian et al 1985, Benson 1985, 1986, Hisazumi et al 1983, 1984a,b, Jocham et al 1983, 1984a, 1985, 1986). The extent and duration of healing and hence the improvement in the survival rates can only be evaluated after long-term observation of the patients. An important factor might be continued effectiveness of carcinogenesis by urine substances. Our first experience with clinical applications of integral photodynamic therapy using a diffusion medium and a special catheter was in October 1984 (Jocham et al 1984). Here we present some comments on the results, which are, however, still preliminary.

Irradiation techniques

Since 1984, various methods have been devised to obtain homogeneous irradiation of the urinary bladder (Table 1) (Benson 1986, Hisazumi et al 1984, Jocham et al 1984b). Figures 1 to 5 illustrate these techniques. A common problem in all the procedures is the exact positioning of the radiation light source and the equipment for intravesical on-line dosimetry. Irradiation techniques that are clinically easy to use are necessary. PDT *per se* does not require any anaesthesia. Instead of a rigid instrument (Kelly et al 1976, Rothauge al 1983, Tsuchiya et al 1983), our research group uses a special catheter which requires, at most, anaesthesia of the mucosa.

Criteria for patient selection and treatment protocol

We only accepted for integral PDT patients with recurrent multifocal bladder tumours who had been recommended to undergo cystectomy after all other local procedures had failed. All the accepted patients had been unsuccessfully treated not only with transurethral electroresection (TUR) and Nd-YAG laser therapies but also with local cytostatic drugs, such as the immunostimmulants Bacillus Calmette-Guérin (BCG), interferon or keyhole limpet haemocyanin. These selection criteria were in accordance with the local ethics commission's decision that PDT could only be applied in 'worst-case' patients.

For all patients, PDT was preceded by an extensive histological evaluation of the extent of the tumour in the bladder. Additional tumours of the urethra

TABLE 1 Technical approaches to integral irradiation of the bladder

Method of irradiation	*Reference*
Bulb shape irradiator on tip of flexible fibre (Fig. 1)	(D. R. Doiron, personal communication)
Moving light pen (Fig. 3)	(Hisazumi et al 1984)
Diffusion medium/special catheter with isotropic light detectors (Fig. 4)	(Jocham et al 1984b)
Fibre tip positioning using special cystoscope with isotropic detectors along the bladder wall (Fig. 5)	(Star et al 1987)

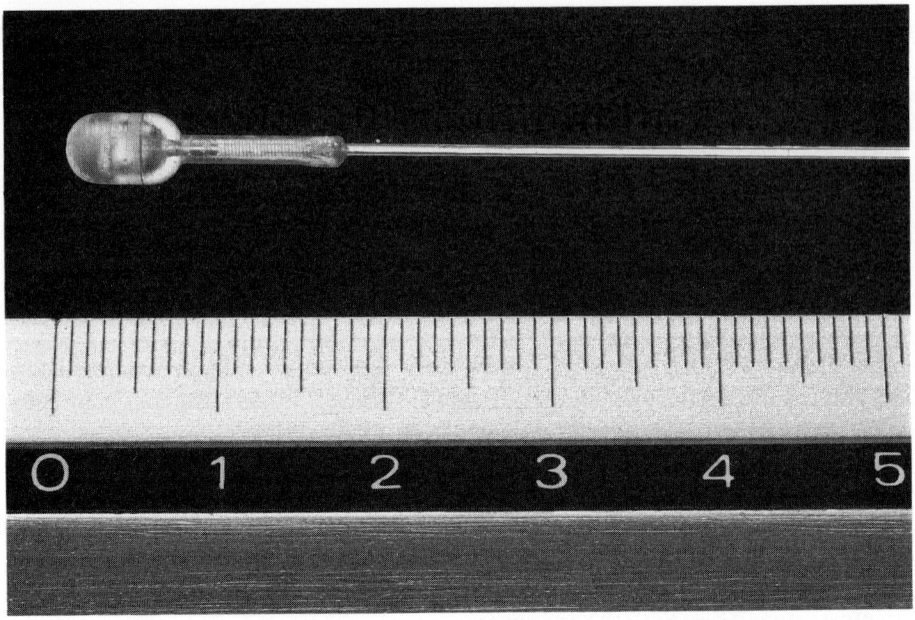

FIG. 1. Isotropic spherical radiator (D. R. Doiron, personal communication) at the tip of a flexible fibre. (Scale in cm.)

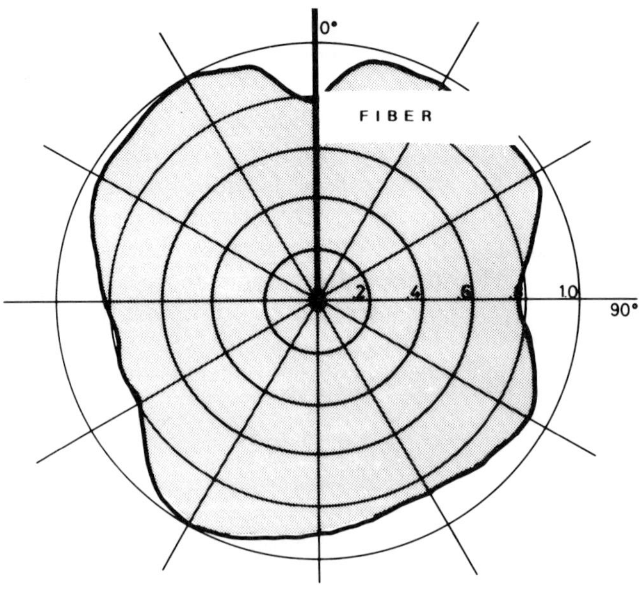

FIG. 2. Radiation pattern of a spherical radiator. Under clinical conditions inhomogeneities are to be expected at the bladder outlet.

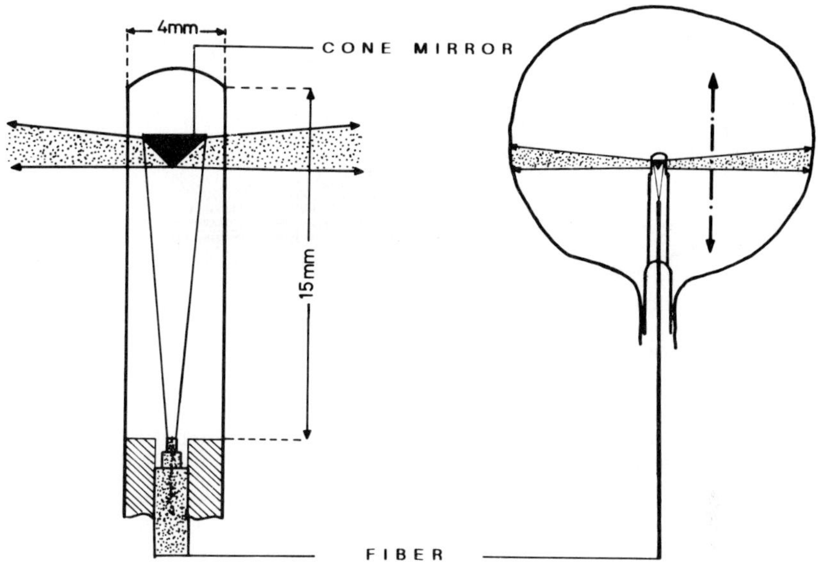

FIG. 3. Computer-aided motor-driven irradiation system with a circular radiation pattern (Hisazumi 1984b). The light beam is moved at various speeds depending on the distance between light source and bladder wall.

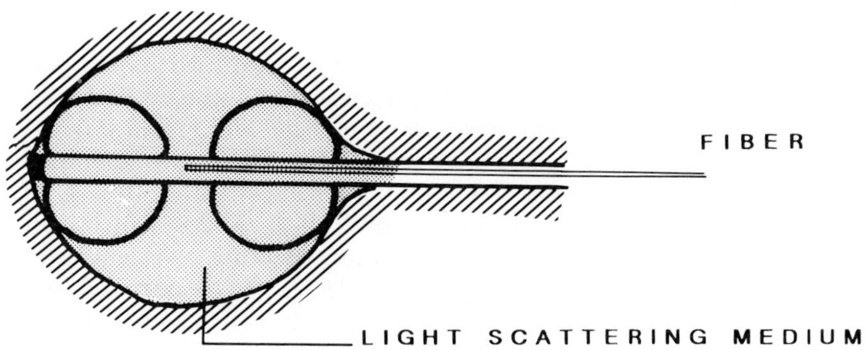

FIG. 4. Integral PDT of the bladder using a light-scattering medium (1:40 Intralipid 10% in NaCl at a bladder volume of 150 ml). The central positioning of the fibre necessary for homogeneous irradiation of the bladder is achieved by using a special catheter (Fa. Rüsch, D-7053 Kerner-Rommelshausen).

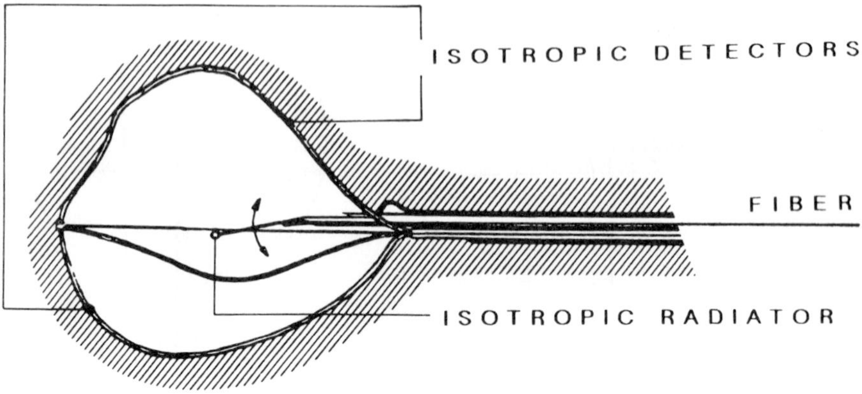

FIG. 5. Movable spherical radiator in a rigid endoscope (Star et al 1987). The central position is monitored by means of various isotropic detectors guided close to the bladder wall.

were excluded by biopsies. Another criterion for selection was that endovesical ultrasonography, computed tomography (CT) and magnetic resonance imaging gave no evidence of metastases, tumours of infiltrating growth or tumours which exceeded the organ. For one patient, PDT was preceded by pelvic staging lymphadenectomy. Exophytic tumour fractions of superficial tumours (TA,T1) were resected separately before PDT to ensure that treatment was required only for histologically confirmed carcinomas with in-situ foci and/or flat TA-lesions. The number of patients with an appropriate history to fulfil all these criteria is very small, according to our experience. Clinical evaluation is therefore considerably restricted. All patients accepted the treatment after extensive counselling about its experimental character.

Intravenous application of the photosensitizer (dihaematoporphyrin ether (DHE) or haematoporphyrin derivative (HpD) at a dose of two to three mg/kg body weight) was performed slowly in the presence of anaesthetists (Berenbaum et al 1982). Protection from light, particularly sunlight, was provided from the moment of application until the end of the time in hospital. Irradiation was done 60 to 72 h (DHE) or 40 to 48 h (HpD) after intravenous injection. The first three patients were irradiated with a power of 2.5 to 2.8 W at the tip of the light guide at a calculated dose of 70 J/cm^2. Because of extensive local effects, the dose was subsequently reduced to 15 J/cm^2 and later changed to 35 J/cm^2. It is difficult to calculate the irradiation dose, and these values may not be correct. The calculation is based on the assumption of a bladder volume of 150–200 cm^3 during irradiation and a 10% light absorption within the diffusion medium (Intralipid 10%) at a dilution of 1:40 in saline solution. A significant source of error in these dose calculations may be back-scatter of laser light towards the diffusion medium from the bladder wall.

Although reliable values of irradiation dose are not available, clinical experience has shown a direct correlation between required therapeutic effects and side-effects. This is true for the symptomatology and bladder capacity. In all patients, permanent bladder irrigation via a transurethral catheter was applied for at least two, and up to nine, days after irradiation. In addition to laboratory analysis, all patients were investigated by renal ultrasonography every three days during their stay in hospital for detection of any urinary obstruction. The first four patients underwent an additional endoscopy under anaesthesia with biopsy six to eight days after irradiation. Now the first routine check-up is not performed until six weeks after irradiation. All but two of the 29 patients presented regularly for the three-monthly check-ups after the first follow-up investigation. This routine includes laboratory analysis, renal ultrasonography and endoscopy of the bladder as well as urinary cytology and many biopsies within the scope of bladder mappings. Once a year a computed tomography of the abdomen is done.

Results

The results presented here refer to 15 patients who were treated by PDT and followed up comprehensively over at least 12 months. The treatment of the first patient was started in October 1984. Thus the longest follow-up is 54 months. Preliminary results show complete tumour destruction for 23 of the 28 tumours (83%). With hindsight, the failures in therapy can be ascribed to technical problems. In all cases we observed side-effects such as pronounced pollakisuria and dysuria that slowly declined and partly disappeared after seven months. In the early phase of PDT extensive bladder mucosal lesions are regularly found. For some patients the whole mucosa sloughs. Although the dose has now been reduced to 35 J/cm^2, the observed capacity reduction varies; in some patients a reduction of more than 50% is found. The extent of bladder shrinkage seems to correlate with the irradiation dose. Such bladder shrinkage is eventually reversed, even after high light doses. In one example, the reduction in initial capacity was from 350 cm^3 to 90 cm^3 after PDT at a dose of 70 J/cm^2, whereas a capacity of 400 cm^3 was detected after six months and throughout the following three years. Only in three of 18 patients was the capacity one year after PDT found to be less than 30% of the initial capacity. Even for these three patients the capacities were still in the normal range, being at least 250 cm^3.

Discussion

In the long term, recurrence of bladder tumours must be expected. For example, despite normal findings during preceding follow-up investigations, tumour recurrence was observed in one patient after 19 months, and in another patient after 37 months. The recurrence in the former was unilocular, and in the latter multilocular. These patients were subjected to transurethral resection and Nd-YAG laser therapy. Five other patients show histologically simple dysplasias without definite evidence of malignancy. As yet, the relevance of these dysplasias to later recurrence is unclear. In all patients there is histological evidence of inflammation, even a long time after treatment. Endoscopic examinations reveal that the bladder mucosa is more easily irritated than normal. Cold biopsies are mostly difficult to perform because of scarring of the bladder wall.

After PDT, transitory distension of the upper urinary tract was observed by ultrasonography in five cases. In two patients this distension receded, and in one it may be attributed to unilateral vesico-ureteral reflux. The other two cases are due to a moderate blockage without a scintigraphically detectable reduction in function.

Conclusions

Even in cases where all other treatments failed, integral PDT enables a complete destruction of superficial bladder carcinomas. However, successful treatment is accompanied by considerable transitory side-effects. In particular, patients suffer cystitis and bladder shrinkage which take a long time to heal.

Tumour recurrences cannot be avoided, even by using local application of PDT. This is probably because of continuing local initiation of carcinogenesis by substances in the urine. Both PDT dosimetry and the tumour selectivity of the photosensitizers need to be improved (Jocham et al 1984). Preliminary indications of successful PDT, even for 'worst-case' patients, justify its further application, but the definite status of this therapy still has to be determined.

Acknowledgement

This work is supported by the Ministry of Research and Technology under grant No. 0706903.

References

Althausen AF, Prout GR, Daly JJ 1976 Non-invasive papillary carcinoma of the bladder associated with carcinoma in situ. J Urol 116:575–580

Baghdassarian R, Wright MW, Vaughn SA, Bern MW, Martin DC, Will AG 1985 The use of lipid emulsion as an intravesical medium to disperse light in the potential treatment of bladder tumors. J Urol 133:126

Benson R Jr, Farrow G, Kinsey J, Cortese D, Zincke H, Utz D 1982 Detection and localization of in situ carcinoma of the bladder with hematoporphyrin derivative. Mayo Clin Proc 57:548–555

Benson RC 1985 Treatment of diffuse transitional cell carcinoma in situ by whole bladder hematoporphyrin derivative photodynamic therapy. J Urol 134:675

Benson RC 1986 Integral photoradiation therapy of multifocal bladder tumors. Eur Urol 12 (suppl 1):47–53
Berenbaum MC, Bonnett R, Scourides PA 1982 In vivo biological activity of the components of hematoporphyrin derivative. Br J Cancer 45:571–581
Hisazumi H, Misaki T, Miyoshi N 1983 Photoradiation therapy of bladder tumors. J Urol 130:685–687
Hisazumi H, Miyoshi N, Naito K, Misaki T 1984a Whole bladder wall photoradiation therapy for carcinoma in situ of the bladder. A preliminary report. J Urol 131:884–887
Hisazumi H, Miyoshi N, Misaki T 1984b A trial manufacture of a motor-driven laser light scattering optic for whole bladder wall irradiation. In: Doiron DR, Gomer CJ (eds) Porphyrin localization and treatment of tumors. Alan R Liss, New York, p 239–247
Jocham D, Staehler G, Chaussy Ch, Hammer C, Löhrs U 1981 Laserbehandlung von Blasentumoren nach Photosensibilisierung mit Hämatoporphyrin-Derivat. Urologe A, Suppl 20:340
Jocham D, Staehler G, Chaussy Ch et al 1983 Integral dye-laser irradiation of photosensitized bladder tumors with the aid of a light scattering medium. In: Doiron DR, Gomer CJ (eds) Alan R Liss, New York, p 249–256
Jocham D, Staehler G, Chaussy Ch, Löhrs U, Unsöld E 1984a Integrale Photoradiotherapie des Blasenkarzinoms nach tumorselektiver Photosensibilisierung mit Haematoporphyrin-Derivat (HpD). Akt Urol 15:109
Jocham D, Schneckenburger H, Unsöld E, Weinsheimer W 1984b Time-resolved laser fluorescence and photobleaching of single cells after photosensitization with hematoporphyrin derivative (HpD). In: Andreoni A, Cubeddu R (eds) Porphyrins in tumor phototherapy. Plenum, New York, p 137
Jocham D, Staehler G, Baumgartner R, Unsöld E 1985 Die integrale photodynamische Therapie beim multifokalen Blasenkarzinom—erste klinische Erfahrungen. Urologe A 24:316–319
Jocham D, Schmiedt E, Baumgartner R, Unsöld E 1986 Integral laser-photodynamic treatment (PDT) of multifocal bladder carcinoma photosensitized by HpD. Eur Urol 12 (suppl 1):43–46
Jocham D, Baumgartner R, Fuchs N, Stepp H 1987 Fluorescence imaging of porphyrin-sensitized bladder tumors. In: Jacobi GH et al (eds) Investigative urology 2. Springer-Verlag, Berlin
Kelly JF, Snell ME 1976 Hematoporphyrin derivative: a possible aid in the diagnosis and therapy of carcinoma of the bladder. J Urol 115:150
Rothauge CF, Röttger P, Kraushaar J, Kracht J, Nöske HD 1983 Die Fototherapie des Blasenkarzinoms im Licht der Histologie. Diagnostik Intensivther 8:17
Star WM, Marijnissen HPA, Jansen H, Keijzer M, van Gemert MJC 1987 Light dosimetry for photodynamic therapy by whole bladder wall irradiation. Photochem Photobiol 46:619–625
Tsuchiya A, Obara U, Miwa M, Oi T, Kato H, Hayata Y 1983 Hematoporphyrin derivative and laser photoradiation in the diagnosis and treatment of bladder cancer. J Urol 130:79

DISCUSSION

Dougherty: The biggest problem in treating bladder cancer with photodynamic therapy (PDT) is dosimetry. Initially, we didn't realize that we couldn't easily locate the centre of the bladder. The first 15 or 20 patients suffered terrible side-effects because we were giving high light doses at the trigone area, which is the worst place for damage to occur. Using ultrasonography we were able to centre the fibre more accurately. We calculate that if we are off centre by 1 cm when using the bulb fibre without any scattering media, the distribution of light changes markedly. If you use scattering media this effect is even more drastic.

Jocham: I agree. We made those measurements by constructing cages surrounded by pig bladder and using isotropic detectors both inside and outside. But in clinical applications we use magnetic resonance imaging to look at the centering of the fibre with the double balloon catheter. We can achieve quite reproducible setting of this system, but we decided to change it because we wished to make treatment even more comfortable. We like to use a light delivery and distribution system which is similar to typical urological procedures, i.e. a smooth catheter that can be easily inserted and gives good conditions for scattering. That enables us to do PDT without anaesthesia.

Carruth: The bladder is not a sphere. Therefore centering is a relatively meaningless concept.

Dougherty: There may not be even light distribution, but that's the best you can do.

Wilson: If one simply calculates the surface area and divides that into the power transmitted down the fibre the answer is completely different from the actual fluence delivered as surface irradiation because of the re-absorption of the back-scattered light in the enclosed volume.

Dougherty: What do you suggest to solve that problem?

Wilson: Star et al (1987) put some isotropic detectors on the bladder wall. That is a good approach; at least it measures the total surface fluence.

Jocham: That's one solution, but Star's method involves procedural problems. Our idea is to have isotropic detectors on the outer surface of the balloon. The catheter itself and the reflections from the inner bladder surface provide sufficient diffusion of light. Nevertheless, for safe application of PDT in the urinary bladder one should use an inner dosimetry during treatment.

Wilson: If you put a detector at the centre of the bladder and know the bladder shape you can do some rough calculations.

Jocham: But, as Dr Carruth mentioned, each patient has a bladder of different shape.

Baer: What is the selectivity of HpD for bladder tumours?

Jocham: In nitrosamine-induced bladder tumours in the rat, which are not really carcinoma-in-situ, the accumulation of HpD or dihaematoporphyrin ether (DHE) in the tumour is only twice that in normal tissue. In infiltrating highly malignant tumours the tumour to normal tissue ratio may increase to 6:1.

Bown: In my research group Abram Pope studied the mechanical properties of rat bladders after PDT, because of your earlier results with irritable bladders. By treating the whole bladder with a fibre inserted through a catheter and roughly centered, he produced necrosis through most of the bladder wall. He then inflated the bladder. A normal untreated bladder started to ooze as the pressure reached about 300 mmHg. It didn't burst, but the fluid diffused through the bladder wall and the pressure couldn't get any higher. When he made a thermal lesion with a laser a slightly higher pressure was reached for a few seconds, but then the bladder burst through the most damaged area of the lesion. To our surprise, the strength of a bladder treated by PDT was almost three times the strength of a normal bladder; we had toughened rather than weakened it. We only studied one point in time after treatment. However, in an equivalent experiment in the colon the strength was maintained at all times after PDT from a few minutes up to several weeks (Barr et al 1986).

Dougherty: Dr Jocham, does PDT treatment induce a lot of fibrosis?

Jocham: We can only refer to two patients. They had scarring mainly within the submucosal area. In our experimental models, using the same dosages as used for patients, with the same possible inherent mistakes, we induced fibrosis in the muscle layer. After PDT we see a wide variety of local reactions that we cannot correlate with any history of inflammation or radiotherapy. Nearly all patients have some scarring and it's very difficult to take biopsies from the bladders.

Bown: Surely the depth of the scarring is a function of the PDT parameters. With lower light and drug doses the scarring doesn't penetrate to the muscle.

Jocham: That is true. We have to consider using other wavelengths and other dosages. We can't be sure of the diagnosis of carcinoma-in-situ. We decided to start with red light therapy to ensure at least some penetration to the depth of the bladder wall. Now that we know we are able to induce good therapeutic effects by PDT we can investigate the effects of adjusting the parameters.

Bown: In our compliance studies, as we put up to 1 ml of saline solution into a normal rat bladder, the pressure required only increased slightly, but in the PDT-treated bladder the pressure increased markedly. Thus the elasticity is destroyed by PDT, but it recovers later.

Lim: Does that cause clinical problems, such as urinary retention?

Jocham: Yes; patients suffer bladder shrinkage. We had to perform a cystectomy for one patient because the almost complete loss of bladder capacity became intolerable. We also see contraction of the orifices, which might explain the dilatation of the upper urinary tract. Nevertheless, we have not needed to use balloon catheter dilatation or set a nephrostomy.

Kato: In your fluorescence detection studies you mentioned the lack of fluorescence in mild urothelial dysplasia and the presence of fluorescence in severe urothelial dysplasia. We obtained the same result in the bronchial tree.

You found a false negative rate of 4%. What causes false negative and false positive diagnoses of carcinoma-in-situ?

Jocham: False positives arise mainly because of inflammation. Using the electronic device we could set a higher threshold for the signal. That would give a greater specificity but lower sensitivity. The false positive biopsies are not a serious clinical problem because the pathology studies tell us that there is only inflammation. However, the false negatives are worrying. Our preliminary experience is based only on animal experiments. In dog bladders we observed a false negative rate of 12%. When we removed and folded the bladders, the same procedure gave a reduced incidence (4%) of false negatives. This occurs because of limitations in the instrumentation.

Kaye: The tumours that you treat are very thick and would be easy to debulk with a mechanical device. The problem with selectivity will not be overcome until a better drug is found. Why don't you use a much shorter wavelength to prevent the problems of the transmural irradiation?

Jocham: We must investigate all the side-effects, but we could not decide to use shorter wavelengths when we had followed only 15 or 20 patients for long periods. The main reason not to change is that we are never sure about the onset and depth of tumour infiltrations. Violet or green light gives a maximum range of 2–3 mm, whereas red light offers more penetration. From the scarring in the submucosa and, from our experimental data, in the muscle layer, we were confident of killing the tumour with red light. Unfortunately, the cost is the risk of severe side-effects.

Kaye: In the animal experiments you affect structures outside the bladder. You are obtaining much deeper effects than are necessary.

Jocham: Different animal models have different characteristics and resistances to PDT. There is very little reaction in dog bladders, for example. It's very difficult to extrapolate from animal experiments to PDT in man.

Dougherty: Dosimetry is more difficult to control with green light than with red light. Dr U. O. Nseyo has treated four patients using 514 nm light. He found that even the tiny amount of blood present in the bladder disrupted the experiment because of its strong absorption at that wavelength.

Moan: When you use balloons to position the fibre, does the pressure of the balloon against the bladder introduce hypoxia?

Jocham: Hypoxia is not a problem. The balloon does not exert much pressure on the wall. The main reason for using a balloon is to enable us to use scattering materials inside it.

Moan: If you were to reduce the drug dose, the treatment would be much less dependent on the exact light dosimetry.

Jocham: Yes, we plan to do that.

References

Barr H, Tralau CJ, Krasner N, Boulos PB, Clark CG, Bown SG 1987 Photodynamic therapy of the normal rat colon with phthalocyanine sensitisation. Br J Cancer 56:111–118

Star WM, Marijnissen HPA, Jansen H, Keijzer M, van Gemert MJC 1987 Light dosimetry for photodynamic therapy by whole bladder wall irradiation. Photochem Photobiol 46:610–624

Photoradiation therapy of brain tumours

Andrew H. Kaye

Department of Neurosurgery, Royal Melbourne Hospital, Melbourne 3050, Victoria, Australia

Abstract. Present methods for treating cerebral glioma are inadequate and the possible benefit of using photosensitization therapy to obtain improved local control of the tumour has been studied in the laboratory and in clinical trials. The biological basis for photoradiation therapy and the laboratory studies and clinical trials using photoradiation to treat cerebral tumours are discussed. Photoradiation therapy results in selective tumour destruction in an intracerebral glioma model with an effect up to 1 cm in depth. Clinical studies using haematoporphyrin derivative and up to 260 J/cm^2 of red light from laser sources indicate that the therapy is well tolerated and may be of value as an adjuvant treatment of cerebral tumours.

1989 Photosensitizing Compounds: their Chemistry, Biology and Clinical Use. Wiley, Chichester (Ciba Foundation Symposium 146) p 209–224

Cerebral tumours are responsible for approximately 2% of all cancer deaths and the high-grade cerebral glioma is the most common cerebral tumour in adults. This malignant tumour invades extensively through the brain and there is no satisfactory treatment. The best treatment available, using surgery, X-irradiation therapy and systemic chemotherapy, results in a median survival of less than one year (Walker et al 1978). Most treatment failures are due to local recurrence of the tumour, which suggests that a more aggressive local therapy, such as photoradiation therapy to the tumour, could be beneficial for treatment of gliomas.

The first investigation of the use of photoradiation therapy on cerebral glioma was made in 1972 when Diamond et al (1972) examined the effect of haematoporphyrin activated by white light on glioma cells *in vitro* and on a transplanted glioma tumour in rats. Photoradiation therapy was first used as a treatment for human gliomas in 1980 (Perria et al 1980).

Laboratory investigation of photoradiation therapy for cerebral tumours

Photoradiation therapy has been used to kill glioma cells *in vitro* and *in vivo* (Diamond et al 1972, Granelli et al 1975, Laws et al 1981) and highly selective uptake of haematoporphyrin derivative (HpD) into animal cerebral tumour models has been demonstrated (Kaye et al 1985, Wharen et al 1983). The optimal characteristics of photosensitizers for brain tumours may not be the same as those for non-central nervous system tumours. Drugs that do not cross the blood–brain barrier, but can enter gliomas, may be the best agents, because the selective localization of the photosensitizer to the brain tumour and its exclusion from the normal brain is probably related to a relative lack of blood–brain barrier within the tumour: we have shown that uptake of HpD into normal brain tissue occurs only into areas known to be outside the blood–brain barrier, or when the blood–brain barrier is disrupted by mannitol (Kaye et al 1985).

We have used the C6 glioma rat model and xenograft mouse model to study phototherapy in the laboratory because the C6 histological pattern closely resembles that of the human glioma.

Photosensitizer uptake and localization

Haematoporphyrin has been shown by fluorescence studies to be selectively localized in cerebral lesions and to be excluded from the normal brain (Wise & Taxdal 1967). In the C6 model, HpD is highly localized to the tumour areas, with slight fluorescence in the brain adjacent to the tumour and no significant fluorescence in the normal brain tissue, except for areas outside the blood–brain barrier (Kaye et al 1985). These findings correlate well with measurements of blood–brain barrier permeability in a rat glioma model which showed markedly increased permeability of the blood–brain barrier within the tumour, with partial preservation of the barrier in the brain adjacent to the tumour (Yamada et al 1982). In this region, the oedematous brain may contain tumour cells that are within a relatively intact blood–brain barrier.

Patchy uptake of HpD into the 9L gliosarcoma rat model has been demonstrated, with maximum fluorescence 24 hours after the administration of HpD (Boggan et al 1984). In the C6 glioma model, maximal and uniform fluorescence occurred throughout the tumour by six hours (Kaye et al 1985). These variations probably reflect histological differences between the tumour models. Using tritiated HpD, Wharen et al (1983) showed that the tumour:brain ratio of HpD four hours after injection into rats bearing ethylnitrosourea-induced central nervous system tumours was twice that after 24 hours. In contrast, maximal uptake of HpD has been reported at 24 hours in extracerebral transplanted tumours (Gomer et al 1982). This difference in the time to maximal fluorescence further suggests that the bases for tumour selectivity of HpD in gliomas and systemic tumours are different; in extracranial tumours, selective retention rather than selective uptake seems to be important.

Studies with tritiated HpD also showed that the amount of photosensitizer localizing in normal brain was extremely small in both animals and humans (Wharen et al 1983). Similarly, HpD administration in the C6 glioma model led to no significant fluorescence in normal brain.

The method of measuring porphyrin uptake is a major limitation in the qualitative and quantitative estimation of the uptake of HpD and Photofrin II. There are problems in using radiolabelled porphyrins; HpD is a complex mixture of porphyrins and studies with tritiated HpD are therefore not entirely convincing because the specific activity or distribution of the ^{3}H label may not be equal in all components. The major problem with fluorescence microscopy of tissue is that this is an indirect method for measuring porphyrin levels *in situ*. At high concentrations of porphyrin an internal quenching effect may occur, producing non-linearity of fluorescence, and the sensitizing and fluorescent components may not be identical. However, in view of the difficulties in obtaining uniformly radiolabelled HpD, fluorescence estimations of uptake remain at present the best option. We have used a porphyrin extraction technique whereby the components of HpD can be extracted from tissue and the amounts determined by a fluorescent spectroscopic assay.

Other porphyrins, such as porphyrin C, have been synthesized in pure form. Porphyrin C has been radiolabelled with ^{35}S, allowing accurate radiotracer studies to be performed. These studies have confirmed a high degree of selective tumour uptake 45 minutes after intravenous administration, with no significant amount of radiolabelled porphyrin in the normal brain and a tumour:brain ratio of 1000:1.

Photoradiation therapy in a laboratory model

Selective tumour kill with sparing of the normal brain can be achieved with photoradiation therapy in the rat C6 intracerebral glioma model (Kaye & Morstyn 1987). With 20 mg/kg HpD and 200 J/cm^2 of 630 nm red light from an argon-pumped rhodamine dye laser, the mean depth of tumour kill was 4.5 mm and in 20% of animals the depth of tumour destruction was greater than 6 mm. Increasing the dose of HpD did not increase the depth of destruction but it increased oedema and both the likelihood of development of necrosis and its extent in normal brain. There was a significantly greater depth of tumour kill (to a maximum of 1 cm) if higher doses of red light were used, but the high doses increased the risk of necrosis in normal brain. Prolonged survival times with photoradiation therapy have been reported in rats with both 9L (Boggan et al 1985) and C6 (Kaye et al 1988) intracerebral tumours.

Provided that sufficient HpD is present within the tumour to mediate a photochemical reaction, the depth of tumour necrosis is largely the result of penetration of red light through the tumour. It is also dependent upon the position of the tumour within the normal brain and light penetration through the surrounding normal brain. Svaasand & Ellingsen (1983) found that the depth of penetration of light in malignant gliomas was usually twice the penetration depth in the surrounding normal tissue. The depth of penetration of red light (defined as the point at which light density is e^{-1} of the starting value) in glioma was between that in adult brain and in neonatal brain—approximately 2.5 mm. The greatest penetration depth was in the most malignant gliomas. Thus the extent of tumour destruction reported in the C6 model is two to three times greater than the reported penetration depth of red light. This additional tumour destruction may result from vascular damage. Experimental studies which support this hypothesis demonstrated evidence of vascular endothelial damage 15 minutes after phototherapy.

The effect of photoradiation on normal brain

Photoradiation therapy has been shown to damage normal brain at laser doses ranging from 330 J/cm^2 to 1382 J/cm^2 (Cheng et al 1984), but in the rat model no effect on normal brain was evident at red light doses of up to 200 J/cm^2 and a dose of 20 mg/kg HpD. Phototoxicity to normal brain may result either from HpD diffusing into the normal brain or from a photochemical reaction involving the sensitizer in small vessels in normal brain, resulting in damage to those vessels and consequent infarction of the brain tissue. However, studies on the entry of HpD into normal brain have been contradictory; some investigators have reported no significant accumulation of HpD in normal brain, except in areas outside the blood–brain barrier (Kaye et al 1985, Wharen et al 1983, Wise & Taxdal 1967) whereas others have reported uptake of HpD (Rounds et al 1982). It is probable that HpD does accumulate in the wall of the blood vessels of normal brain but it is unlikely that it penetrates into normal brain parenchyma.

Thermal effect

A heating effect occurs in tumours treated with photoradiation therapy. Irradiation with 1200 mW led to a temperature rise of up to 7 °C after one minute. This could be avoided by irrigating with normal saline at room temperature. The role of hyperthermia in the tumoricidal effect in phototherapy is disputed, but tumour destruction has been demonstrated in the absence of significant heating of the tissue. Perhaps hyperthermia could be used in association with phototherapy to enhance the photodynamic cell kill.

The laboratory findings thus show that: photoradiation therapy does produce selective tumour kill; the dose of HpD and the dose of light are critical in achieving selective tumour kill; the effective depth of kill is 4 to 10 mm; tumour kill occurs independently of a hyperthermia response; and, in limited studies, the survival of animals with cerebral tumours is improved.

Photoradiation therapy in clinical neurosurgery

The first attempts at photodynamic treatment of human gliomas were reported in 1980 (Perria et al 1980, Perria 1981). Since 1978, a total of 86 patients with

TABLE 1 Reported photoradiation treatments of cerebral tumours

Study	No. of patients	Power (W)	Light dose: Total dose (J)	Light dose: Total dose to tumour (J/cm^2)
Perria et al (1980), Perria (1981)	9	0.025	n.a.	0.9–9
McCulloch et al (1984)	16	0.280–0.460	1620–2520	n.a.
Laws et al (1981)	5	0.250–0.400	540–1440	n.a.
Wharen et al (1983)	3	n.a.	n.a.	180[a]
Muller & Wilson (1985)	8	n.a.	439–3888	8–68
Kaye et al (1988)[b]	50	0.75–4	3360–10413	72–260

n.a., information not available.
[a]Derived from 100 mW/cm^2 for 30 min delivered by xenon arc lamp.
[b]Published in part (Kaye et al 1987, 1988).

cerebral tumours have been treated with photoradiation therapy (Table 1). Most of the tumours treated have been the adult high-grade glioma (glioblastoma) because this is the most common adult cerebral tumour and the present treatment is unsatisfactory. The light sources used were a xenon arc lamp (Wharen et al 1983), helium neon laser (Perria et al 1980), argon-pumped dye laser (Laws et al 1981, McCulloch et al 1984, Muller & Wilson 1985) or gold metal vapour laser (Kaye et al 1987). The results of the initial clinical studies of photoradiation therapy have been disappointing. However, the tumours treated were often recurrent gliomas and the doses of light used were up to 100 times lower than those used for systemic tumours, partly because of fear of the side-effects of high doses, especially when combined with X-irradiation therapy, and partly because of a lack of powerful light sources.

The use of photoradiation therapy to treat brain tumours involves a number of technical problems, including the choice of sensitizer, its dose and the time of its administration, and the selection of the irradiating system and light dose. The possible effects of preoperative steroid administration, and potential postoperative and postphototherapy complications, including cerebral oedema, skin sensitization and interaction with X-irradiation therapy, must also be considered.

Sensitizers

HpD and Photofrin II have been used as sensitizers in neurosurgical trials. They have been administered intravenously; direct injection into the tumour has been advocated (Kostron et al 1986), but this would seem to be inappropriate because the basic concept of phototherapy is its 'selectivity' for tumour rather than normal brain. Any adjuvant therapy for cerebral gliomas should be directed towards an improved tumoricidal effect on the spreading edge of the tumour rather than on the central mass, which can usually be resected by standard surgical techniques. Although an intra-tumour injection would provide high doses of sensitizer within the tumour mass it would be less likely to provide a selectivity for the tumour cells spreading out into the normal and often vital area of brain.

The dose of HpD administered intravenously has varied from 2.5 to 5 mg/kg and Photofrin II has been administered at 2.5 mg/kg. In the series of patients with malignant cerebral tumours treated at the Royal Melbourne Hospital–Ludwig Institute for Cancer Research (RMH–LICR), 5 mg/kg of HpD has been given intravenously 24 hours before surgery because the laboratory data indicate this to be the optimal time. In other trials the sensitizer has been administered up to three days preoperatively.

Tumour excision

The position and size of the tumour determine the extent to which it can be resected before phototherapy. Because of the limited depth of penetration of the irradiating light, the most favorable tumours for phototherapy are those in which a radical 'de-bulking' operation can be performed, leaving as little residual tumour as possible. A radical de-bulking procedure was performed for all patients in the RMH–LICR series, with the aid of the Cusa Cavitron (Cooper Laboratories Inc, Stamford, Connecticut, USA).

Light-irradiating systems

Laser systems are the most efficient method of delivering red light. They have the advantages that: a single known wavelength of light is produced; the light output can be accurately measured; the light can be administered selectively to parts of the tumour cavity; deep cavities can be effectively irradiated; and high doses of light can be used. The argon-pumped rhodamine dye laser produces continuous light but an inefficient coupling mechanism between the lasers results in the output of only relatively small doses of light. This laser is relatively immobile and requires a dedicated area for its operation. The gold metal vapour laser is portable, produces a much higher light intensity at 627.8 nm and has the theoretical advantage of being a source of 'pulsed' rather than continuous light. Thus it produces very intense peaks of light approximately 10 000 times per second, which may improve sensitizer activation and increase the depth of penetration.

In the RMH–LICR series laser light was generated using an argon-pumped rhodamine dye laser (15 patients) or a gold metal vapour laser (35 patients) and was delivered through a flat-cut quartz fibre. The power at the fibre tip was in the range of 0.7–4.0 W. The fibre was attached to a brain retractor arm and the tumour bed was irradiated evenly for between 43 and 94 minutes (median time 60 minutes). The diffusing agent used was 0.5% Intralipid. The temperatures were monitored on the surface of the brain and in the brain deep to the irradiated surface and were kept below 37 °C by irrigation, although it is realized that hyperthermia might have a synergistic effect with photoradiation therapy. The total time for surgery, including the photoradiation therapy, varied from three to 5.5 hours.

Photoradiation dose

We used a light intensity of up to 4.0 W at the fibre tip and a tumour dose of up to 260 J/cm^2 with the gold metal vapour laser, although the laser power used in previous studies varied from 12.5 mW (McCulloch et al 1984) to 400 mW (Laws et al 1981) and the dose of light varied from 0.5 J/cm^2 to 9 J/cm^2 (Perria et al 1980), 8 to 68 J/cm^2 (Muller & Wilson 1985) and 100 mW/cm^2 (Wharen et al 1983). Too few patients were treated and the follow-up was too short in our series for us to say whether the higher doses of pulsed light ($>$120 J/cm^2) delivered by the gold metal vapour laser were more effective than the lower doses of continuous light previously used, although a trend favouring higher doses of light in gold metal vapour lasers was observed (Fig. 1). Muller & Wilson (1985) used an inflatable balloon to help achieve an even distribution of light. However, this method would not be appropriate for patients with uneven cavities or where resections result in a flat surface after the tumour excision.

Side-effects

Increased cerebral oedema has been reported after photoradiation therapy (McCulloch et al 1984). However, this was not evident in our 50 patients and our laboratory studies indicate that increased cerebral oedema only occurs at doses greater than 20 mg/kg HpD and 200 J/cm^2 of red light.

Skin photosensitization results from HpD infusion and we have instructed patients to remain out of direct sunlight for four weeks. The patients are then asked to gradually increase their daily exposure to the sun. The longest period of significant sensitization has been 16 weeks, with a median period of seven weeks. It is possible that lower doses of Photofrin II would produce less skin photosensitization. Although carotene-containing substances might theoretically diminish the degree of skin sensitization, there have been no clinical reports substantiating their use.

The possible interaction of postoperative radiotherapy with a porphyrin sensitizer has been a concern. Although it is improbable that the photosensitizer does cross an intact blood–brain barrier, the HpD may remain in small vessels for some time. Consequently, it has been our policy not to commence

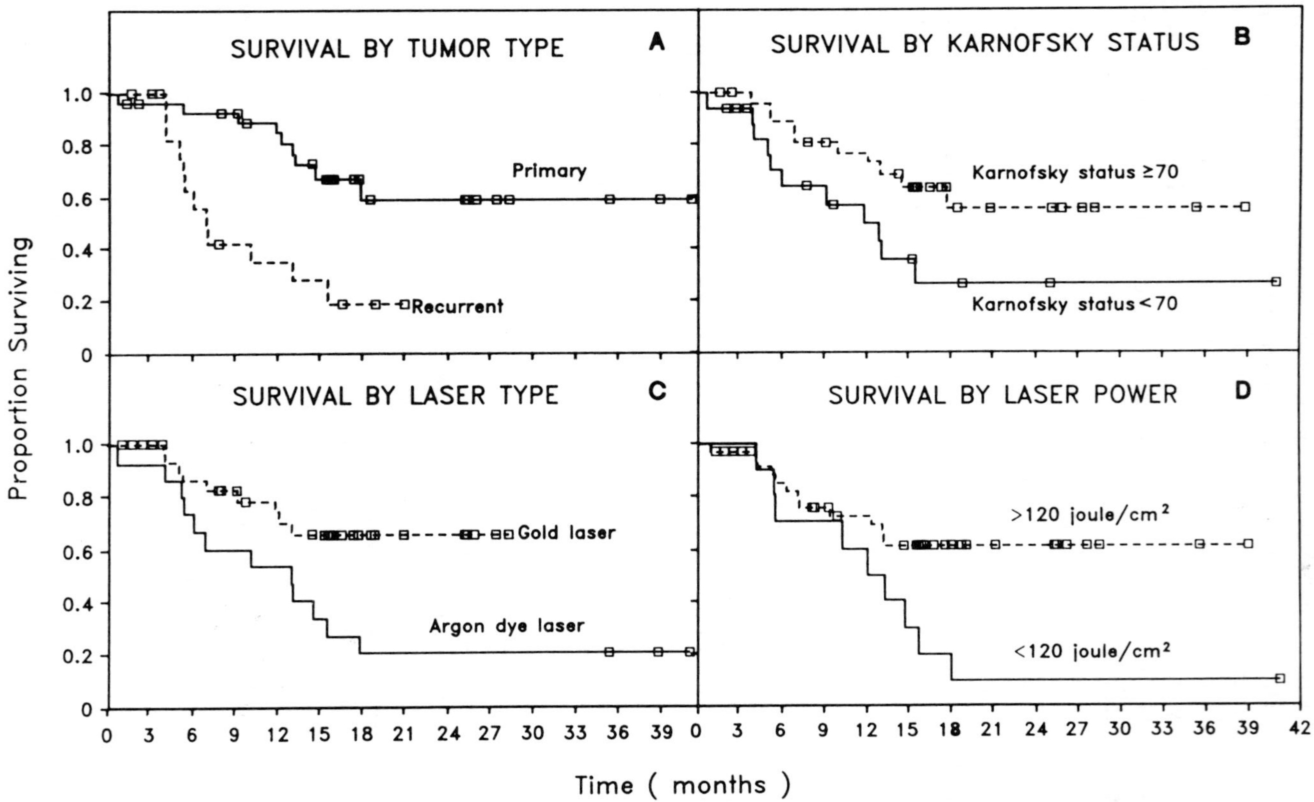

FIG. 1. Survival curves of patients with high-grade gliomas after surgery, photoradiation therapy and X-irradiation. The survival times are compared for patients with primarily treated non-recurrent and recurrent gliomas (A), patients with Karnofsky performance status of <70 and $\geqslant 70$ (B), patients treated with the gold metal vapour laser and argon-pumped rhodamine dye laser (C), and for treatment with photoradiation doses of <120 or >120 J/cm^2 (D).

radiotherapy until four weeks after the administration of photoradiation therapy. In 29 patients treated with conventional radiotherapy after phototherapy there has been no increase in the toxicity of postoperative X-irradiation therapy.

While it is generally thought that there are no interactions between chemotherapy and phototherapy, this has not been exhaustively investigated. *In vitro*, anthracyclines and HpD each appeared to interfere with the cellular uptake of the other (Bohmer et al 1987).

Results of photoradiation therapy

Although the reported series all have a relatively short follow-up period and the early results were disappointing, there have been some long-term survivors (Laws et al 1981, McCulloch et al 1984). We have treated 50 patients with malignant cerebral tumours, forty-nine of whom had high-grade gliomas (24 glioblastoma, 18 recurrent glioblastoma, six anaplastic astrocytoma, one recurrent anaplastic astrocytoma; Table 2). One patient had a metastatic tumour from the lung. All patients who presented with recurrent gliomas had previously received radiotherapy, and patients with new gliomas underwent postoperative radiotherapy (45 Gy in 20 divided doses).

Twelve of the 19 patients with recurrent gliomas have had further recurrent tumours and have died. The median survival after these tumours appeared was six months. The remaining patients are clinically disease free after a median follow-up time of five months.

Nine of the 30 patients with primarily treated gliomas had a clinical and CT scan recurrence between three and 19 months after therapy. A further patient died of an acute myocardial infarction 15 days postoperatively. The other patients have been followed clinically for up to 41 months with no clinical evidence of tumour recurrence (Fig. 1). The number of patients in this study is too small and the duration of follow-up too short to draw conclusions about the efficacy of the therapy.

There were no direct complications from the photoradiation therapy in this study: there was no significant increase in neurological deficit following surgery and photoradiation therapy; no evidence of increased cerebral oedema; and no increase in the toxicity of postoperative X-irradiation therapy.

Future applications

Further follow-up is necessary for us to determine whether photoradiation therapy is a beneficial adjuvant to surgery and X-irradiation. A problem with testing this is that tumour responses to photoradiation are difficult to measure. Because the depth of tumour kill is likely to be only 0.5 to 1.0 cm, photoradiation therapy probably has little to offer as the sole treatment for large tumours. A properly instituted randomized trial for cerebral gliomas is required. However, as

TABLE 2 Characteristics of patients with recurrent and primarily treated gliomas

	Recurrent gliomas	*Primarily treated gliomas*
No. of cases	19	30
Median age (years)	42	51
No. of males	11	19
Anaplastic astrocytoma	1	6
Glioblastoma multiforme	18	24
Duration of symptoms (months): <6	16	18
>6	3	12
Karnofsky performance status score: <70	7	13
>70	12	17
Type of laser: Argon-pumped dye	6	9
Gold metal vapour	13	21
Dose of laser light (J/cm^2): <120	4	6
>120	15	24
Laser power: <1.5 W	6	10
>1.5 W	13	20
Cases with recurrence	12	9
Deaths	12	9

such a trial would probably require approximately 200 patients to be statistically significant, there are obvious problems with logistics and organization.

At present the major limiting clinical factors in treating central nervous system tumours by photoradiation therapy are the selectivity of tumour uptake, particularly in the region of the brain adjacent to the tumour, and the penetration power of the porphyrin-activating system, such as the laser light.

A number of new sensitizers are being tested in the laboratory, including porphyrin C (Scourides et al 1985), *meso*-tetrahydroxyphenylporphyrins, phthalocyanines (Ben-Hur & Rosenthal 1985) and rhodamine 123 (Powers et al 1986). Phthalocyanines have a strong absorption band in the red region of the visible spectrum and they should theoretically be more efficiently activated by red laser light to kill tumour cells than HpD. Rhodamine 123 is a lipophilic cationic dye that is specifically taken up by mitochondria of living cells, and Powers et al (1986) have shown that it is selectively retained by cultured glioma cells. However, rhodamine 123 is only activated by blue-green light, which is significantly less able to penetrate through most biological tissues than the longer wavelengths; this would severely limit its clinical application.

Laser systems producing light of the appropriate wavelength to activate the new sensitizers will need to be developed. The use of higher powers to increase the depth of tumour kill will be limited by the tolerance of the normal adjacent brain. Metal vapour lasers have improved portability, high power and the theoretical advantage of the 'pulsed' dose. The excimer laser systems, which have the potential for developing very high powers (up to 120 W), are not yet sufficiently developed for clinical use. Advanced technology solid-state lasers pumped by diode lasers have the potential for producing a wide range of wavelengths at high power, but the cost will be a major hurdle for their development for medical use.

The use of phototherapy in combination with stereotactic equipment is an exciting possibility for the treatment of small, deep-seated, previously unresectable tumours, but a major technical problem is to shield the fibre tip and prevent local charring and excessive heating. It will be possible to use photoradiation therapy to treat brain tumours other than the high-grade gliomas. Low-grade gliomas, large pituitary tumours, inaccessible meningiomas, pineal tumours and posterior fossa medulloblastomas and ependymomas are all likely to concentrate photosensitizer selectively. Photoradiation therapy may be of benefit in these tumours where total resection is not possible and an adjuvant therapy to treat the residual tumour is indicated.

The future application of photoradiation therapy will depend upon the success of clinical trials of the treatment of cerebral tumours. However, the photosensitizers, irradiating systems and techniques of delivery must be improved so that the technique can become a useful adjuvant in the therapy of a wide range of central nervous system tumours. The future advances will thus depend on a better understanding of the basic sciences associated with photoradiation—especially photophysics and sensitizer chemistry.

Acknowledgements

I thank Diane Edwin for typing the manuscript. The laboratory studies were undertaken with the aid of grants from the National Health and Medical Research Council, the Anti-Cancer Council of Victoria, Stroke Research Foundation, Royal Australasian College of Surgeons and the Royal Melbourne Hospital Victor Hurley Medical Research Fund.

References

Ben-Hur E, Rosenthal I 1985 Photosensitised inactivation of Chinese hamster cells by phthalocyanines. Photochem Photobiol 42:129–133

Boggan JE, Walter R, Edwards MSB et al 1984 Distribution of hematoporphyrin derivative in rat 9L gliosarcoma brain tumour analysed by digital video fluorescence microscopy. J Neurosurg 61:1113–1119

Boggan JE, Bolger C, Edwards MSB 1985 The effect of hematoporphyrin derivative photoradiation therapy on survival in the rat 9L gliosarcoma brain tumor model. J Neurosurg 63:917–921

Bohmer RM, Hoffmann K, Morstyn G 1987 Hematoporphyrin derivative and anthracyclines mutually inhibit cellular uptake and toxicity. Cancer Chemother Pharmacol 20:16–20

Cheng M-K, McKean J, Boisvert D, Tulip J, Mielke BW 1984 Effects of photoradiation therapy on normal rat brain. Neurosurgery 15:804–810

Diamond I, Granelli SG, McDonagh AF, Nielson S, Wilson CB, Jaenicke R 1972 Photodynamic therapy of malignant tumors. Lancet 2:1175–1177

Gomer CJ, Rucker N, Mark C, Benedict WF, Murphee AL 1982 Tissue distribution of H-hematoporphyrin derivative in athymic 'nude' mice heterotransplanted with human retinoblastoma. Invest Ophthalmol Visual Sci 22:118–120

Granelli SG, Diamond I, McDonagh AF, Wilson CB, Neilson SL 1975 Photochemotherapy of glioma cells by visible light and hematoporphyrin. Cancer Res 35:2567–2570

Kaye AH, Morstyn G 1987 Photoradiation therapy causing selective tumor kill in a rat glioma model. Neurosurgery 20:408–415

Kaye AH, Morstyn G, Ashcroft RG 1985 The uptake and retention of hematoporphyrin derivative in an in vivo/in vitro model of cerebral glioma. Neurosurgery 17:883–890

Kaye AH, Morstyn G, Brownbill D 1987 Adjuvant high dose photoradiation therapy in the treatment of cerebral glioma: a phase I-II study. J Neurosurg 67:500–505

Kaye AH, Morstyn G, Apuzzo M 1988 Photoradiation therapy and its potential in the management of neurological tumors—a review. J Neurosurg 68:1–14

Kostron H, Bellnier D, Lin C-W, Schwartz MR, Martuza R 1986 Distribution, retention, and phototoxicity of hematoporphyrin derivative in a rat glioma. Intraneoplastic versus intraperitoneal injection. J Neurosurg 64: 768–774

Laws ER Jr, Cortese DA, Kinsey JH, Eagan RT, Anderson RE 1981 Photoradiation therapy in the treatment of malignant brain tumors: a phase 1 (feasibility) study. Neurosurgery 9:672–678

McCulloch GAJ, Forbes IJ, Lee See K, Cowled PA, Jacka FJ, Ward AD 1984 Phototherapy in malignant brain tumours. In: Doiron DR, Gomer CJ (eds) Porphyrin localization and treatment of tumours. Alan R Liss, New York, p 709–717

Muller PJ, Wilson BC 1985 Photodynamic therapy: cavitary photoillumination of malignant cerebral tumours using a laser coupled inflatable balloon. Can J Neurol Sci 12:371–373

Perria C 1981 Photodynamic therapy in human gliomas by hematoporphyrin and He-Ne laser. IRCS (Int Res Commun Syst) Med Sci (Cancer) 9:57–58

Perria C, Capuzzo T, Cavagnaro G et al 1980 First attempts at the photodynamic treatment of human gliomas. J Neurosurg 24:119–129

Powers SK, Pribil S, Gillespie Y, Watkins P 1986 Laser photochemotherapy of rhodamine-123 sensitized human glioma cells in vitro. J Neurosurg 64:918–923

Rounds DE, Jacques S, Shelden CH, Shaller CA, Olson RS 1982 Development of a protocol for photoradiation therapy of malignant brain tumors: I. Photosensitization of normal brain tissue with hematoporphyrin derivative. Neurosurgery 11:500–505

Scourides PA, Bohmer RM, Henderson RW, Farragher M, Morstyn G 1985 N,N′-diacetyl porphyrin C: its purification, ^{14}C labeling and its photosensitizing properties. In: Jori G, Perria C (eds) Photodynamic therapy of tumors and other diseases. Libreria Progetto, Padua, p 10–12

Svaasand LO, Ellingsen R 1983 Optical properties of human brain. Photochem Photobiol 38:292–299

Walker MD, Alexander E Jr, Hunt WC et al 1978 Evaluation of BCNU and/or radiotherapy in the treatment of anaplastic gliomas. A cooperative clinical trial. J Neurosurg 49:333–343

Wharen RE Jr, Anderson RE, Laws ER Jr 1983 Quantitation of hematoporphyrin derivative in human gliomas, experimental central nervous system tumors, and normal tissues. Neurosurgery 12:446–450

Wise BL, Taxdal DR 1967 Studies of the blood–brain barrier utilizing hematoporphyrin. Brain Res 4:387–389

Yamada K, Wshio Y, Hayakawa T, Kato A, Yamada N, Mogami H 1982 Quantitative autoradiographic measurements of blood–brain barrier permeability in the rat glioma model. J Neurosurg 57:394–398

DISCUSSION

Dougherty: The key to success is the ability to destroy the nests of cells outside the main tumour. In the animal studies you have shown that the treatment goes beyond the tumour. Is that beyond the oedematous area or within it?

Kaye: The infiltrating tumour is usually within the oedematous area. We do destroy the tumour cells outside the main tumour in our C6 glioma rat model, but do we kill the animals? It's not easy to perform a craniotomy on a rat, remove the tumour, give irradiation and close up the excision. The improved survival is statistically significant but not dramatic. Jim Boggan used the 9L gliosarcoma model, which is a highly aggressive tumour, and also reported improved survival, but it was only just statistically significant. The problem is that the tumour cells are infiltrating still further.

Dougherty: How do you apply the light? Is the fibre in the tissue?

Kaye: No. We use a flat quartz fibre. If you put it into the tissue, the tissue chars, particularly at those doses; the power at the fibre tip is up to 4 W.

Carruth: You use a straight-cut fibre onto a diffusing medium in the brain cavity. Have you considered using a diffusing bulb on the fibre for more uniform irradiation of the cavity?

Kaye: We tried that; both bulbs we used exploded.

Dougherty: The cavities involved have large volumes. By the time the light reaches the walls it's totally diffuse, whatever the source looked like.

Berenbaum: Although you and others have shown by fluorescence and radiolabelling that there is very little haematoporphyrin derivative (HpD) in normal brain and much more in brain tumours, we found in mice that the small amount that entered normal brain was lethal. It seemed to be associated with the vascular endothelium, and photodynamic damage to the endothelium caused cerebral oedema and death (Berenbaum et al 1986). That is why we felt that PDT was not a viable proposition for brain tumours. Your work suggests that we were unnecessarily pessimistic. However, I am puzzled by your conclusion that sensitizers more selective than HpD are needed. HpD appears to be perfectly selective, because you don't see damage to normal brain, or even cerebral oedema, in patients.

Kaye: We need a drug that gets into the infiltrating tumour cells away from the tumour mass and that can be activated with longer wavelength light which penetrates further.

Morgan: You also have the problem of skin photosensitivity.

Kaye: This doesn't seem a very important issue when one is treating serious diseases such as brain tumours. To a thirty-year-old woman with two children the necessity of staying out of the sun for a month is at most a nuisance.

We would trade off skin photosensitivity if we could use higher doses of a drug and retain tumour selectivity. If we increase the dose of HpD beyond that which we now use clinically, we observe necrosis. In the rodent studies, if we use 20 mg/kg HpD and less than 240 J/cm^2, we get selective tumour kill. At higher doses the drug enters normal brain tissue and necrosis results.

Berenbaum: Using a similar dose in mice and only 10 J/cm^2, we kill the mice within a few hours, with or without craniotomies.

Kaye: Mice are very difficult to work with. We do uptake studies in mice to measure selectivity, but we use rats for the treatment studies. It's difficult to operate on mice atraumatically; performance of craniotomy without injury to the underlying brain is a major technical feat. D. E. Rounds irradiated through the overlying skull, but there is an associated heat effect. That is a very important issue.

Berenbaum: We use 10 J/cm^2 which we can deliver in 30 s. A significant temperature rise is not likely in that time.

Kaye: No, probably not.

Wilson: The reports on oedema are not inconsistent. We have measured intracranial pressure after PDT in a series of patients. Compared with a control surgery-only group, the intracranial pressure is doubled after PDT, which shows we are damaging some tissue. As Dr Kaye points out, you can control that effect clinically. I suspect that severe oedema and an increase in intracranial pressure kills the mice.

Berenbaum: But Dr Kaye has shown an undamaged rat brain with a necrotic tumour underneath (Kaye & Morstyn 1987).

Kaye: That depends on the dose. With 20 mg/kg HpD and less than 240 J/cm^2, we don't detect any effect on normal brain by light microscopy. When we increase the dose we start affecting normal brain, for the reasons you outlined; drug gets into the endothelium wall and causes a vascular effect. It then leaks through the endothelium and damage to normal brain results.

Dougherty: Normal brain is one of the most opaque tissues, mainly because of the high light scattering. What is the maximum improvement in light penetration that we anticipate in brain tissue?

Wilson: That is very hard to estimate. Normal brain is one of the few tissues in which penetration continues to increase dramatically as you go from 650 to 800 nm. We have not measured the penetration depth beyond 800 nm, but it is still increasing.

Carruth: What depths of penetration occur with 630 nm and 800 nm light?

Wilson: The penetration depth doubles in going from 630 to 800 nm. In pig brain the penetration increases from about 2 to 4 mm.

Bown: You can produce a biological effect down to about three penetration depths.

Dougherty: It is important not to equate penetration depth and the depth of biological response; they are not the same.

Wilson: We and Dr Kaye have both shown substantial clinical responses to PDT. Therefore I presume that doubling the penetration depth would be significant.

Dougherty: The problem with using wavelengths longer than 800 nm is that eventually the water absorption becomes important.

Wilson: That doesn't arise until 900 nm.

Dougherty: Do you treat your patients with dexamethasone before the HpD is injected?

Kaye: These patients present with raised intracranial pressure and often travel a long way for PDT. First they are prescribed dexamethasone or steroids by a general practitioner, and they are controlled on steroids before surgery.

Dougherty: Do steroids inhibit the uptake of the photosensitizing drug?

Kaye: No, they do not.

Henderson: Most of the drugs we study would be selectively taken up in tumours because of the blood–brain barrier in normal brain.

Kaye: As you say, virtually anything will enter the main tumour. We are aiming at a drug that will selectively enter islets of tumour cells outside the main tumour mass.

Henderson: Are those isolated tumour cells supplied by diffusion of nutrients or do they already have their own vasculature? The very small tumours will not take up the drug if the blood–brain barrier still exists.

Kaye: That is the key issue. From our monoclonal antibody studies we seem to be affecting those cells within the oedematous region adjacent to the main tumour. It's crucial to get a drug that will be selectively taken up into islets of cells at a distance from the tumour mass. If we try to measure the blood–brain barrier using standard techniques we shall not see any difference, but the microenvironment of the area around the islets of tumour may in fact have broken down.

Dougherty: Jim Boggan and others also showed an uptake of the drug in the oedematous area adjacent to the tumour.

Wilson: We have been treating patients with much lower light doses. Your Figure 1 shows longer survival with light doses greater than 120 J/cm^2. Are those matched groups, or was there bias towards the higher doses being given to smaller tumours?

Kaye: Not at all. Initially we used 70 J/cm^2, which we considered to be a fairly low dose, because of warnings about cerebral oedema and the harmful effects of radiotherapy. Since then we have increased the dose. High-grade glioblastomas are almost always the same size at the time of diagnosis, about 10^{11} cells, at which stage they are big enough to cause increased intracranial pressure.

Wilson: Did you observe complete resolution of the CT-enhancing lesion at lower light doses? We treated six patients at much lower doses. Although survival was not increased, we did observe a clinical response (Muller & Wilson 1987).

Kaye: We see clinical remission and CT remission at lower light doses but until we do a Phase III matched study we can't be sure that phototherapy is the cause.

Wilson: We treated two patients with cystic lesions. No resection was done; the cysts were drained and PDT was given to relieve the intracranial pressure. That treatment showed CT responses.

Kaye: There's no doubt that phototherapy will kill the tumour, but whether it kills enough to be clinically significant is the question.

References

Berenbaum MC, Hall GW, Hoyes AD 1986 Cerebral photosensitization by haematoporphyrin derivative. Evidence for an endothelial site of action. Br J Cancer 53:81–89

Kaye AH, Morstyn G 1987 Photoradiation therapy causing selective tumour kill in a rat glioma model. Neurosurgery (Baltimore) 20:408–415

Muller PJ, Wilson BC 1987 Photodynamic therapy of malignant primary brain tumours: clinical effects, post operative ICP, and light penetration of the brain. Photochem Photobiol 46:929–935

Final general discussion

Drug and light doses for photodynamic therapy

van den Bergh: The doses of red light reported in PDT with porphyrins vary by several orders of magnitude. It is important to be able to measure such doses accurately. This can be achieved reasonably well at the skin or in an organ of regular geometry, such as the oesophagus, but is very difficult in organs with complex geometries, such as the lungs.

In Lausanne we have studied the problem of light distribution in patients treated for cancer of the oesophagus. We use a light distributor consisting of a cylindrical bulb at the end of a plastic tube which stretches the oesophagus to a diameter of about 18 mm over a length of about 8 cm. An optical fibre conducts light from the laser to a diffusing tube at the centre of the bulb. We measure the light intensity with an *in situ* photodiode and the temperature with a thermocouple.

The distribution of light emerging from the bulb is purposely not homogeneous. We test this bulb in the oesophagus of a freshly slaughtered calf and measure the light distribution on the outside of the oesophagus as well as in its wall. However, even if we start with a homogeneous distribution of light from the bulb, the distribution in the oesophageal wall is not homogeneous because of light scattering in the tissue. In treating bronchial tumours we have induced stenosis in a few patients. This can be avoided by using hemicylindrical rather than cylindrical light distributors. The same holds for the oesophagus.

Professor Jocham has used fluorescence methods to detect bladder and lung tumours. Fluorescence endoscopy is extremely useful for detecting small tumours when white light endoscopy does not show their presence or position. We built an apparatus that is in principle a bronchoscope and allows alternate white light illumination of the lungs, so that we can see what we are doing, and blue light irradiation from a krypton laser, which excites fluorescence from the dye in the tumour. The white light image goes to a CCD camera. The fluorescent light image is split into two colours (red and green) which are amplified by an image intensifier so that the red autofluorescence can be subtracted from the red tumour fluorescence for improved contrast. Professor Jocham excites at two wavelengths consecutively, observes the fluorescent images only in the red region (600–700 nm), and then subtracts these two red images. We excite at one wavelength (410 nm) and detect simultaneously at two wavelengths for autofluorescence subtraction.

For efficient detection of early tumours one needs a substance that is selectively taken up by the tumour and fluoresces strongly. One of our approaches (in collaboration with the Institute of Biochemistry at UNIL) has been to tag monoclonal antibodies with an average of ten molecules of fluorescein per antibody molecule. We compare the fluorescence intensities of the porphyrins and the fluorescein-tagged antibodies in several tumour-bearing mouse organs. The intensity of fluorescence from a tumour in a mouse 2–3 days after injection of antibodies tagged with 1 μg fluorescein is 35 times that in a comparable mouse injected with 100 μg HpD. Excitation and observation wavelengths are 488 and 520 nm respectively for fluorescein and 410 and 690 nm for HpD. The contrast of tumour to muscle fluorescence is about 10:1 in the best cases with HpD and at least 50:1 with the fluorescein-tagged antibody. This does not necessarily imply that the use of antibodies is the solution to our detection problems in the clinic, because extrapolation from mice to humans is difficult. But we should be looking for selective fluorescent dyes to improve detection of early cancer, as well as searching for new therapeutic dyes.

A further issue is the measurement of tumour depth before treatment. We have had problems with fistula in the oesophagus and should therefore like to know the depth of the tumour before we start PDT. If a shallow tumour containing HpD is irradiated with blue light, which does not penetrate far, fluorescence of intensity 'A' is emitted. Green light would penetrate further and cause fluorescence of intensity 'B'. There is very little porphyrin below the superficial tumour in the normal tissue and hence virtually no fluorescence comes from that region. For a deep tumour, blue light irradiation again produces fluorescence of intensity 'A'. However, the green light penetrates further, and much more fluorescence emanates from the deeper areas of the tumour which contain substantial amounts of the porphyrin dye. Thus, the ratio of fluorescence intensity induced at the two wavelengths depends on the tumour depth and can be used for depth measurement.

Dougherty: The fluorescence ratio for the fluorescein-tagged monoclonal antibody and for HpD, measured to be 35, is almost exactly the same as the relative fluorescence signals for fluorescein and HpD. The monoclonal antibody may not be relevant; the uptake may be the same in each case.

van den Bergh: Fluorescein alone is mostly eliminated from the body within a few hours. We make fluorescence measurements up to three days after i.v. injection. We did control experiments with fluorescein tagged with non-specific antibodies and with pure fluorescein.

Henderson: Are you going to make a monoclonal antibody for each patient?

van den Bergh: We hope to be able to reach a large proportion of lung tumours with a few different monoclonal antibodies. We are testing that.

Carruth: How does a patient with carcinoma-in-situ in the oesophagus present?

van den Bergh: If, for example, a large primary tumour is discovered in the mouth, it may be easy to remove surgically with a good chance of local cure, but the specialist will check the entire upper respiratory and digestive tracts for small secondary tumours that might cause major long-term medical problems. Most of this tumour detection (except in lungs) is done with toluidine blue, which is a simple and effective way of marking early cancer.

Carruth: That is similar to the situation where repeat bronchoscopy is performed for patients with laryngeal cancers because they have a high risk of a second primary tumour. You are only picking up the second primary in patients with advanced cancer elsewhere.

Bown: You said it was important to deliver the correct light dose. But in the majority of cases we don't know what light dose we want.

van den Bergh: That is still a problem. There is probably no single correct light dose for PDT with HpD. For early tumours in the ear, nose and throat region we deliver about 100 mW/cm^2 of 630 nm light over a 20 min period three days after i.v. injection of 3 mg/kg HpD. In future we may try to reduce this HpD dose to about 1 mg/kg.

Dougherty: We have been studying the effects of varying the drug and light doses in patients with various types of cancer. It's a tedious and difficult procedure requiring many patients for useful results. It is clear that by careful choice of Photofrin II dose we can deliver unlimited light to the tumour and avoid non-repairable damage in normal tissues. One incentive for lowering the drug dose was the possibility of a reduction in skin photosensitivity, but it is almost impossible to gather and sort out the necessary information because we don't see patients frequently enough.

In a variety of patients with either metastatic breast cancer or basal cell carcinoma, treatment with the usual dose of Photofrin II (2 mg/kg) results in a MED (the minimum effective light dose to induce slight erythema) of less than 10 J/cm^2 with 630 nm light. With more than 35 J/cm^2 we get necrosis of the skin. When we reduce the drug dose we see better results. With 1 mg/kg we induce a MED (630 nm) of about 20 J/cm^2, and with even less drug the MED is greater than 30 or 40 J/cm^2. We must ensure that as we reduce the dose we retain reasonable tumour responses, but we are trying to use as little drug as possible.

Lim: How long after injection was the MED measured?

Dougherty: This was usually done at the same time as the tumour treatment—two or three days after injection.

Lim: Do you then irradiate and wait for another day to evaluate the treatment?

Dougherty: Yes. Ninety-five percent of patients with basal cell carcinoma can be cured with conventional methods, but they tend to suffer recurrences and sometimes they can't be cured. For example, one patient had a long history of basal cell carcinoma and a large tumour on the end of his nose which was no longer treatable by standard methods. It was suggested that he had extensive surgical removal as an alternative. We gave him 1 mg/kg Photofrin II and treated the tumour with a light dose of 200 J/cm^2. Two years later biopsy shows no tumour in the areas treated. With a dose of 2 mg/kg, a light dose above 35 J/cm^2 would have resulted in necrosis of healthy skin. In a patch test on normal tissue

Chairman's summing-up

T. J. Dougherty

Division of Radiation Biology, Department of Radiation Medicine, Roswell Park Memorial Institute, 666 Elm Street, Buffalo, NY 14263, USA

Haematoporphyrin derivative (HpD) activated by red light was first reported to be effective in eradicating superficial skin tumours in 1979. Since then several thousand cancer patients, many with far advanced disease, have been treated by photodynamic therapy (PDT). In the short time since its inception a considerable amount of science has built up around this treatment, involving multidisciplinary approaches in chemistry, biology, physics and engineering. Numerous new photosensitizers are being examined in experimental animal systems. These drugs are often better defined chemically than HpD (or its relatively purified form, Photofrin II), a mixture of oligomeric esters and ethers of haematoporphyrin and its dehydration products. In particular we are searching for drugs that are retained in malignant tissue and that absorb at long wavelengths.

Examples of photosensitizers discussed in this symposium include phthalocyanines (both sulphonated and non-sulphonated), naphthalocyanines, and various porphyrin-type compounds (chlorins and bacteriochlorins, purpurins), many of which are potentially clinically useful. The biological responses of tumours and normal tissues to PDT have been considered, particularly in relation to direct and indirect effects of tissue destruction. The effects of light in tissue were discussed in terms of improving penetration—for example, by using drugs absorbing at longer wavelength (700–800 nm) than porphyrins ($\approx$630 nm). However, it has been noted that the best that can be expected is to approximately double the effective penetration of materials of the HpD type.

Improvement in penetration could be particularly important in the attempt to control primary brain tumours such as glioblastomas which form small nests of cells outside the bulk of the tumour. Promising preliminary data on the treatment of such patients with HpD have been presented in this symposium. We have also heard about more than five years' experience in treating early-stage inoperable lung cancer in Japan and superficial bladder cancer in Germany. Two unique uses of photosensitizers have been described for the treatment of psoriasis and of cutaneous T cell lymphoma. The relationship of the response to phototherapy to immunologically induced effects has been presented and discussed.

Regulatory approval in Europe, Canada and the USA for PDT using Photofrin II is expected within the next three years. This will make this new treatment available to a larger number of cancer specialists, in whose hands its ultimate utility will be determined.

Index of contributors

Non-participating co-authors are indicated by asterisks. Entries in bold type indicate papers; other entries refer to discussion contributions

Indexes compiled by Liza Weinkove

Subject index